The Experience of Science

A NEW PERSPECTIVE
FOR LABORATORY TEACHING

O. Roger Anderson

TEACHERS COLLEGE PRESS
Teachers College, Columbia University
New York & London

© 1976 by Teachers College, Columbia University

Manufactured in the United States of America

Library of Congress Cataloging in Publication Data:

Anderson, O   Roger, 1937-
    The experience of science.

    (Studies in science education)
    Bibliography: p.
    Includes index.
    1.   Science — Study and teaching.   2.   Educational
psychology.   3.   Science — Experiments.   I.   Title.
Q181.A63   507'.1   75-37967
ISBN 0-8077-2490-4
ISBN 0-8077-2489-0 pbk.

FOREWORD

Science teaching is a unique human undertaking. It involves teachers with students in a field where an attempt is made to be "objective" and not allow an individual's prejudices and proclivities to unduly influence observations and operations. A scientist and a student of science may "want" certain results very badly, but it is part of the discipline of science that he should be prepared to see, record, and communicate results that he may have hoped he would not find. On the other hand, students, teachers, scientists, and the general public can become deeply involved in science. The phenomena that are studied can be extremely fascinating, individuals can be inextricably identified with various approaches to problems, and everyone has crucial stakes in many of the results of science. Thus, science is an enterprise in which we are disciplined to try to be "objective"; but it is also an enterprise in which humans individually and collectively become deeply involved. It is this complex human undertaking to which the science teacher introduces young children and into which he guides young people as they explore wider and deeper.

The teacher of science does have to be scientist-philosopher-psychologist, and a teacher probably cannot succeed unless he is a bit of each. Students can be completely blocked if they do not even know some of the questions to ask; and the teacher with some background experience in science can help students begin to see more clearly the phenomena they study. But knowing how to investigate and how to organize the results of their investigations is also important; and to help students achieve such understandings the teacher must also be part philosopher. Moreover, in this book at least there is a linking of experiences in science with values and ethical considerations. Finally, a major aim of science instruction is of course to nurture the student's intellectual development. To do this effectively, it is essential to understand the general nature of such development, to recognize where a student may be in his development, and to plan science experiences that will use all the resources and imagination at his command.

There is a need for new perspectives for science teaching. There is considerable evidence that young people are not being attracted as much to the sciences as might be expected. After all, there are probably more professional opportunities opened up through study in the sciences than in many other academic areas. More importantly, many young people are not being intrigued by their explorations in science, and this is unfortunate both for them and for science. The dedicated teacher of science is intrigued and captivated by science; he is perplexed when others are not equally curious and as eager to inquire.

It is especially important that we develop new perspectives for laboratory instruction in science. It is a widely accepted credo that experiences in the laboratory are essential to effective science programs. Occasionally this tenet is questioned, however, and, although laboratory experiences survive, the answers offered critics are not overwhelmingly impressive. Certainly laboratory work is not as efficient and effective a mode of instruction as it might be, and it is high time that we take a hard look, with a fresh perspective, at learning in the laboratory.

In this book, Professor Anderson presents a new perspective that links scientist, sciencing, and science. A three-dimensional model for science teaching is suggested that deals with the philosophical orientation of the individual toward natural and social environments, the psychological orientation of the individual toward scientific inquiry processes, and a set of scientific inquiry processes. The methods that a teacher can use to apply this model are discussed with special attention to laboratory instruction. A series of practical laboratory experiences, classified according to the model, are described as exemplars of some of the categories in the model. These exemplars are suggestive of similar experiences in various science disciplines.

The ideas that are discussed in this book can be used to help students achieve greater intellectual growth. The nature of this intellectual development and the kinds of experiences that foster it are described in detail. Certainly, a book that has such a potentiality is worthy of careful consideration.

Willard J. Jacobson
Series Editor

PREFACE

The cumulated body of knowledge in science has grown to massive proportions in recent years, and few people are certain as to the appropriate content to be included in precollege and early college science curricula. It is more clear now than ever before that we must select carefully what kinds of information will be presented and with what kinds of orientation. As a youth I was attracted to the field of science by the orderliness of its store of knowledge and by the tantalizing precision and rapidity with which new knowledge was acquired. This burgeoning characteristic of science, however, has made it a formidable discipline in the view of some modern young people. The immensity of the knowledge structure and sometimes the apparent misuse of this knowledge to the detriment of society have dissuaded some from studying science.

New perspectives are needed to make science teaching more consonant with modern social and scientific ways of thought. The problem is all the more apparent in designing appropriate science laboratory experiences. The laboratory experience, when properly organized, provides a rare opportunity for young people to gain skills in gathering, ordering, and explaining sensory experiences — an opportunity that few other school activities can offer.

Through a careful mixture, in teaching, of demonstration and exposition of scientific processes with analysis of the psychological mental processes used in scientific reasoning, young people can acquire greater confidence in their ability to deal rationally with empirical problems. Science is a uniquely human endeavor that has influenced many facets of modern life. These influences have spread not only to technology and medicine but even to the work of artists who by the medium they use or the theme they convey clearly exhibit that science has left its mark. All too little attention has been given to the human expression that is evident in science as a discipline.

The purpose of this book is to bring a new perspective to the teaching of modern science in the laboratory. I submit evidence that the laboratory experience can be a unique learning experience that enhances the student's awareness of the orderliness of man's interpretations of nature, the creativity of man's mind in explaining the natural environment, and his own ability to grow in rational ways of thinking about natural phenomena. I present a new model of science and the teaching of science. This model combines psychological processes of thinking with more traditional dimensions of science data processing. Moreover, the model presented here admits science as a human endeavor and thus allows for moral analysis of the outcomes and processes of science and also for psychological analyses of how scientists inquire.

This is a book about teaching, and most of what I say here will bear a clear mark of relevancy to teaching science more effectively to young people. An effective teacher is as much concerned about the psychological growth of his students as he is about the subject matter to be taught. A genuine enthusiasm for one's subject matter can indeed do much to enhance student interest in learning. If, however, little mature concern is shown for the student in mastering the discipline, one can be certain that there will be some interpersonal conflict in the classroom. Many of us who teach *do* have a genuine interest in the student as a scholar, but sometimes we lack the systematic knowledge of human behavior that we need to help students mature intellectually. In some cases more experienced teachers have found solutions to these problems and would like some formal explanations for the processes they have found to work so effectively.

This book is intended primarily for high school and college teachers who would like to have a formal understanding of science teaching as it occurs in the laboratory. I trust that this book will provide some systematic ways of thinking about science and the teaching of science that will be useful to you in helping students to grow intellectually and aesthetically through science laboratory experiences.

There are several people who should be acknowledged for their assistance in the preparation of this book. Mr. Thomas Jacobson assisted in the preparation of laboratory experiences as presented in Chapter Five. His devotion in helping me develop these exemplars of science laboratory teaching and his advice about their applicability from the perspective of a high school student were greatly appreciated. I am deeply grateful to my students and colleagues who have challenged me to think about the science laboratory from the perspective of teaching science. One of my doctoral students, Mr. Walter Lowell, made helpful comments about the practicality of some of the ideas presented. Mrs. Karl Lorenz assisted in typing and in proofreading of the manuscript.

O.R.A.

CONTENTS

FIGURES

CHARTS

THE EXPERIENCE
OF SCIENCE

One

A NEED FOR NEW PERSPECTIVES

Events in recent decades have convinced many informed individuals that we cannot separate the products of society from the human processes that produced them. Each accomplishment of society – whether for the common good or, as it sometimes turns out, for the common detriment – bears the unique human signature, as it were, of those who contributed to the product. The artifacts of society are not merely its utensils, dwellings, and tangible art forms. The intangible products of the scholarly disciplines bear the unique imprint of human origin as surely as does a piece of sculptured clay or a chiseled stone. It is not psychologically satisfying to separate the intellectual artifacts of the scholarly disciplines from their source – namely, a collective human endeavor. At times it appears we have tried to impose this artificial separation between science and the intellectual processes of persons who participate in it. We have denied the vital link between science and the matrix of human endeavor that conceived it and sustained it. This separation can seem all the more convenient when the applications of scientific findings appear to be culturally sterile, socially bothersome, or at worst socially or environmentally destructive.

It is useful to have a conceptual framework for the discussion of laboratory teaching. A conceptual analysis of the field of science is presented here in terms that are relevant to the teaching of science and with an orientation that is consistent with the total perspective developed in this book.

Some observers dissect the field of science into three components: (1) the *scientist:* (2) *sciencing:* and (3) the cumulative product of both called *science.* In this triad, a *scientist* is a person with a social role, one who is

1

committed to a life of investigation, to accumulating knowledge, and to possessing a keen sense of prediction about natural phenomena. *Sciencing* is the application of inquiry methods to probe the enigmatic properties of nature that, either for esoteric or for societies' commonly cherished reasons, drive the scientist forward in his work. Then there is *science* – a body of knowledge and methodologies that stand as the cumulative product of the scientist's use of scientific processes.

Thus far, no great travesty has been committed by so defining these three components. The travesty appears when the components are disarticulated. It should be clear that in a coherent view of the field of science, all three components are interconnected. Each component bears a dynamic relationship to its neighbor whereby each influences the other and in turn is influenced by it. The dynamic equilibrium of influences among the various components is represented in Figure 1 by a set of forward- and backward-directed arrows indicating the associations among the components.

$$\text{SCIENTIST} \underset{(4)}{\overset{(1)}{\rightleftharpoons}} \text{SCIENCING} \underset{(3)}{\overset{(2)}{\rightleftharpoons}} \text{SCIENCE (Product)}$$

Figure 1.   **The dynamic coupling of the triad in the scientific enterprise.**   The scientist, who as a person exhibits value judgments, makes decisions about research objectives, and sets paradigms to guide research, is not separated from the product of his enterprise (science). To the extent that we can recognize a methodology applied to a research problem (sciencing), it links the scientist to the product of his endeavor.

The scientist's role as a decision-maker in defining research objectives and selecting theoretical models to guide research is intimately related to the kinds of methodologies he applies in sciencing. There is a system of ethics, involving criteria of quality and appropriateness of methodology selection, that inheres in the role of being a scientist. Hence: a forward-directed arrow from scientist to sciencing. In complementary fashion, the application of methodologies of science to a problem situation generates new questions about the ethics of the research under way, the suitability of the research objective, and necessary corrections in procedure that in due course need to be considered. Moreover, the applied methodologies channel the scientist into certain areas of sensory data acquisition and set the boundaries on the kind of data he can collect. Thus, there is a backward-directed arrow from sciencing to scientist.

Sciencing is also coupled to science as a product. The application of science methodologies to a problem yields new information, either as additional data about existing theories or as evidence for need of new approaches to insoluble problems within the existing body of methodologies. The information gained, moreover, sometimes supplements or adds to current cumulative knowledge or challenges already accepted theories and necessitates new conceptualizations. A forward-directed arrow is therefore shown from sciencing to science. As a complementary response, the cumulated knowledge and theoretical explanations of science, extant in the body of literature and shared among the members of the community of scientists, suggest new ways of applying methodologies and indicate new objectives suitable for use with established methodologies. Thus there is a backward-directed effect of science on sciencing, as shown by the reverse arrow.

The dynamic coupling of the triad components to one another in this model clearly means that the many aspects of science as a scholarly discipline are embedded in a matrix of human endeavor. An illustration may help to clarify this perspective.

Let us consider the case of a physical scientist. We will assume he is a chemist studying organic molecules. He suspects that the molecular structure of a compound can be elucidated by application of spectrophotometric techniques. Earlier studies of similar compounds have shown that such methods yield sufficient data to make valid interpretations of molecular structure. He therefore selects a spectrophotometric method (arrow 1 in the figure). The scientist makes a decision about use of a methodology in terms of its appropriateness to the objective. In making this decision the scientist has applied some rational scientific logic to the body of knowledge already gained and made deductions about the appropriate methodology to apply to the new situation. Therefore in one sense he has already made one mental circuit through the triad in reaching a decision about the new laboratory methodology to be employed. The scientist has performed an act of thinking that moves through arrows 1 and 2, to science, and then back to the scientist, through arrows 3 and 4. In order to understand how this passage has been accomplished we must define *sciencing* rather broadly, but not unreasonably, to include the intellectual as well as the physical operations that a scientist uses to generate new data and make rational interpretations.

Sciencing includes manipulatory research operations and logical analyses. In some cases when the scientist is actively engaged in gathering sensory data from the natural environment, he is sciencing in a physical operational way. In such a case he is using combined intellectual and laboratory or field techniques to gain new information. This is the sense in which we normally think of the laboratory scientist as being engaged in sciencing. In other cases, sciencing consists of analyzing prior-gained information cumulated in the body of knowledge contained within *science* as a product. This is a use of sciencing as a logical operation. Thus, when a scientist makes a decision to analyze certain available methodologies (e.g., chemistry analytic techniques) by reviewing the literature, he makes a mental passage through sciencing to the corpus of knowledge in science. This is an intellectual operation represented by moving through arrows 1 and 2. When the scientist has made an analysis of the prevailing techniques of chemical analysis, he must then decide whether they are adequate or whether he needs to develop a new approach. This is a passage back from science to the scientist as decision-maker, following along the sequence of arrows 3 and 4. In review: the scientist has made a decision to analyze the literature in a research problem area, and has done so (arrows 1 and 2); from this analysis, the scientist is led to make additional decisions about the kinds of methodologies he will apply, as the operation returns to him as a decision-maker (arrows 3 and 4).

In reaching a preliminary decision about the kind of methodology to apply, the scientist has completed a mental circuit through the triad. This is the kind of circuit that most theoretical scientists apply. They do not generate new data by directly examining the natural environment, but use previously gained information to generate new concepts and theories. Other scientists are laboratory or field-based scientists who apply physical operations or methodologies as a means of gaining new data. Obviously they also may and often do travel the mental circuit as well.

When the scientist actually makes a choice of methodologies and applies them to analyze a physical object or a natural event, then he is using sciencing as a physical operation of gathering data. Hence, he again moves through arrow 1 to sciencing. In the example we are using, the chemist may decide to use a spectrophotometric technique. The application of the technique in actual practice either supports or negates the original assumption by the scientist about its appropriateness in this situation. Arrow 4, directed back from sciencing to the scientist, represents this feedback evidence. As a result of the application of this technique, assuming it is appropriate, new knowledge is obtained and increases the total store of information in science (arrow 2). Moreover, some of the unique properties of the compound discovered through use of this technique may suggest a greater range of applicability of spectrophotometric techniques than heretofore realized. This feedback effect is shown by arrow 3. The range of methodologies used in future research can be extended to the extent allowed by the new knowledge.

I have purposefully chosen a linear model, as presented in the figure, rather than a triangular one. In a triangular model, the three components of the science enterprise are placed at the corners of a triangle and the sets of arrows connect the corners. This means that in addition to the set of arrows in the linear model, there would be a set connecting science to scientist directly. This is not acceptable in my view, since the body of knowledge here called science is interpreted and available to the scientist as a decision-maker only as he applies certain logical rules of sciencing to make passage back and forth. Sciencing thus serves as a critical bridge between scientist and science.

In some people's perspective, the triad is disarticulated into binary units — scientist and sciencing, or sciencing and science. The otherwise integrated sequence of components is separated into independent pairs. By segregating scientist from science through separation of the triad into two artificially exclusive dyads, the chain of influence between scientist as a decision-maker and science (the product of his enterprise) is severed. This disarticulation insulates the scientist as a person from the product of his endeavor, providing a comfortable but questionable way of granting immunity to the scientist against responsibility for his methodologies used and discoveries made.

Clearly, we must recognize that the scientist is responsible for his professional behavior — and perhaps more so than most people, since his professional preparation has conferred upon him a degree of critical and creative thought not commonly acquired in the usual course of experience. At this point let me be quite clear: I do not suggest that the recognition of responsibility as stated above should be used as a shackle for creative scientific inquiry. Indeed, recognition of responsibility for research outcomes should clarify the significant role of the scientist in making ethical and substantive-relevant decisions about research objectives and appropriate research methodologies. Interpolating a sense of sciencing between the role of a scientist as a person and his accomplishments is defensible as a tool to help one grapple with the question of what constitutes the discipline of science. But it is an unfortunate inclusion if it becomes an opaque passage that obscures, instead of revealing, the close-coupling between the scientist and his research outcomes. The more we can understand the human intellectual processes that accompany, if indeed they do not drive, scientific endeavors, the better we can appreciate the unique human qualities that make science a common legacy of our culture. Moreover, such a view indicates the clear logic we must use if the products of the scientist are to be used prudently. The scientist in my view cannot be held responsible for the destructive applications made with his discoveries. Every piece of matter in our environment has the potential to be a destructive weapon. The point I am arguing is that science is a human enterprise and is a product, in many ways, of human decision-making. This fact if accepted suggests that we must use clear logic and prudence in the interpretation of scientific findings and in their technological applications to ensure the best benefits for society.

In this book I present a new perspective on the field of science, one that clearly links scientist, sciencing, and science (as a product) to one another. It is a psychological view that encompasses the emotional (aesthetic and motivational) and rational components of science experiences, thus clearly labeling science as a human endeavor. This is a book on the teaching of science, and it has direct relevancy at many points to teaching young people. The focus is upon high school and early college science instruction.

The work is particularly concerned with teaching in the laboratory. It is concerned with helping teachers better understand how science can be taught as a human endeavor.

To teach science as a human endeavor has a threefold implication. It means to convey the following understandings: (1) Science is a product of a highly advanced society. It permeates modern life activities and influences many aspects of modern thought. Understanding of science as a human enterprise provides a particularly clear view of the present state of our civilization. (2) Science is a human enterprise that provides unique opportunities both to develop confidence in rational thinking and to allow for emotional responses to discoveries and creativity, but also to clearly differentiate the role of emotion from that of logic in those processes. (3) Science has an impact upon the social and natural environments. The sometimes destructive experiments which modern nuclear science has set upon the natural environment is one example of massive impact. More clearly in recent years we have come to understand that the technological applications of science often have unplanned side effects deleterious to our social and natural environments. A wise control of scientific applications is indicated.

All of these aspects of science will be evident in this book. However, I will place primary emphasis on the second one. In the discussions much attention will be devoted to the problem of helping young people better understand the orderly nature of scientific thought, the nature of scientific investigations, and the aesthetic dimensions of science that they as learners can come to appreciate through laboratory experiences.

In recent years, science curricula have emphasized the methodology of science, reflecting the widespread interest generated in modern methods of scientific investigation, as opposed to the knowledge store of science. These are known as inquiry-approach or science-process-approach curricula. There are differences in science content and psychological dimensions among these curricula. For purposes of generic classification, however, I will refer to all of them as inquiry curricula. Those that are clearly inquiry-oriented emphasize open-ended laboratory experiences where the student is allowed to generate explanations for disequilibrating or otherwise problem-generating observations. In a thoroughgoing inquiry lesson, the student may actually perform a miniature scientific investigation that in many respects requires creativity and novelty of thought equivalent to that of a scientist. I applaud the dynamic and active view of science that this approach reflects. It is a perspective one hopes will remain, with true fidelity to its purposes, as a flexible component of the science laboratory curriculum. However, to the extent that the inquiry approach focuses on the mechanics of scientific inquiry to the exclusion of other psychological factors in science learning, it is not sufficiently comprehensive.

The excessive use of open-ended experiences in teaching laboratory science can have deleterious effects unless the students are properly prepared for the experience. The frustration aroused by repetitive exposure to problem situations that are far too demanding for a reasonable solution can hardly contribute to a positive view of science or indeed to an appreciation of science as an orderly and creative thought process. Inquiry should be used in appropriate balance with other methodologies of teaching and with due consideration for the scholastic maturity and readiness of the students for the experience. I will include inquiry concepts as a part of the new perspective presented in Chapter Three.

I have chosen to bring a new perspective to bear on the teaching of science by considering the laboratory as a center of science teaching. The laboratory setting is unique among the various environments that young people encounter in their schooling; it is a peculiar possession of science instruction. It is a place where students can investigate natural phenomena in an immediate or first-hand experience and apply various cognitive skills toward an interpretation of these phenomena. To be sure, various disciplines in the school curriculum offer experiences that facilitate development of thinking skills, but science is a unique discipline in which students can learn to apply systematic thought to the analysis of measurable natural phenomena as presented in immediate sensory experience.

This, then, is the rationale for the laboratory experience as a human enterprise. It is an experience in which the student can learn to appreciate the scientist as a person fulfilling a role in society; he can engage in "sciencing," with its characteristic logical acivities, while engaged generally, as a student, in developing more systematic and orderly ways of thought. Some students — those who can go further — may even achieve a degree of creativity and insight that approximates the achievement of an accomplished scientist.

As a corollary to this view, we must recognize that students possess varying capacities to assimilate science process skills. These differences in preparation and perhaps in basic ability should be considered in planning curricula, to allow for a range of intellectual accomplishments suitable to the scholastic maturity of the student. In other words, the school laboratory experience should match experiences to the developmental level of the student and allow for growth toward more sophisticated behavior. In this book I present a model that allows for such differentiation of accomplishment. The model of science processes presented in Chapter Three will, I trust, clarify how students of various abilities can acquire science skills at various levels commensurate with their preparation.

The school laboratory experience should also serve a psychological positive transfer function. What one learns in the school laboratory should provide generalized ways of thinking that are not limited to interpretations of science phenomena. Reasoning skills and basic levels of confidence acquired through systematic analysis of science problems can be transferred to other spheres, including daily problems in living. This is not given as a cause to ignore the substance of science as a carefully ordered body of knowledge or as a systematic method of inquiry. Rather it is a liberal educational view of the traditional substance of science. The substance of science indeed must be presented in a thorough manner if the psychological interpretations of its meaning are to have significance.

Finally, the person with experience in a laboratory-based science should perceive natural phenomena with a different perspective than one who has not. His view is an appreciation of orderliness and pattern in man's interpretation of phenomena that all modern liberally educated persons are entitled to enjoy.

Two

CONTEMPORARY PERSPECTIVES

New perspectives are very often based upon those that have preceded them. That will be the case here, as we frame a new perspective on laboratory science teaching. This chapter presents a summary of some modern perspectives on science inquiry. The purpose is to find models that help us understand the differences that exist between scientific interpretations and other ways of thinking about reality.

Modern philosophers have contributed much to our understanding of the processes of science. Philosophers are not in agreement, however, about the way we interpret experience. Some believe that reality is given in the natural environment as a set of clearly delineated entities and that man discovers these entities and explains their characteristics. Other philosophers begin with the assumption that there are no immediately given entities in the environment. They assume, rather, that the environment presents certain kinds of sensory experiences (stimuli without a necessarily predetermined organization) and that man imposes certain categories on, and invents explanations for, these otherwise minimally organized sensory inputs.

The latter viewpoint is particularly useful in understanding scientific theory-building, and it has received considerable attention by curriculum writers. As we discuss this view and examine several models, it will be helpful to keep the following brief summary in mind: *Man's ability to order and classify sensory data allows him to impose categories upon natural phenomena and hence determines the way he conceives of reality at any point in time. What we call reality is the product of a dynamic interplay between sensory stimuli and man's capacity to organize and explain these sensations.*

Reality as we know it today may change tomorrow as man reinterprets sensory experience.

Henry Margenau has contributed a brilliant analysis of science that can help us to understand this way of explaining reality.[1] He views science as a process of constructing a set of abstractions that represent sensory experience. These abstractions (ways of explaining experience) are developed in response to sensory stimulation (visual, auditory, tactile, taste, and olfactory stimuli). Margenau's model is represented graphically in Figure 2. Nature is the source of all perceivable stimuli that impinge upon man. As a source of perceivable physical stimulation it is called the P field. The P plane represents man's immediate response to sensory stimulation. The P plane is the interface of man as a thinking organism with the environment as a source of stimuli. It is the most elementary or fundamental level of receiving sensory stimuli. It is a level of information-processing where man recognizes that "something is there" in the environment. He senses an "it." He posits a separation of an object from the surrounding area.

Consider, as an example, a person viewing a distant landscape. The light rays reflected from the environment are collected by his eye and project an image on the retina. This immediate form of sensation, the retinal image, consists of a range of colors and degrees of darkness and light. From this optical image the observer senses an object in the field of vision; he posits that a branching vertical form is present and separates it from the surrounding field. He calls the branching form "a tree." This simple example illustrates how a name ("tree") is associated with a separated part of visual sensation that man posits to exist. A P-plane-level interpretation therefore is assigning names and categories to immediately sensed experiences. Experience is further elaborated by building more abstract constructs such as classificatory categories and non-tangible aspects of experience such as force, inertia, or energy.

The C field is the set of abstractions (interpretations) that man has created to represent sensory data. Each abstract representation of experience is called a *construct;* in the figure, constructs are shown as small circles. Constructs are generated by experiences with physical phenomena, as is shown by a line directed to a construct from the P plane. Some constructs represent very immediate kinds of experience. They are the names, for example, of perceived objects and groups of objects. Other constructs are derived from these basic ones by logical thought processes. This is shown in the figure by lines connecting the circles. The constructs that are most closely linked to immediate sensory experience are shown closest to the P plane, whereas those that are more abstract and logically derived from other constructs are located further in from the P plane. All constructs that have some logical connection

[1.] H. Margenau, *The Nature of Physical Reality* (New York: McGraw-Hill, 1959).

to perceptual experience are contained within the larger circle. Note that all of these constructs can be linked through a network of connecting lines back to the *P* plane. Many logical deductions can be made about phenomena we expect to occur based on prior interpretations of experience. Consider again the example of sensing an object we call a tree growing on a mass of earth we have called a hill. From this very immediate construct we can logically follow into more abstract constructs. We can posit that the tree has mass, that it uses energy to keep itself structurally sound, and that it is anchored in the hill to maintain itself against wind currents. None of these statements is a direct representation of immediate sensory data; rather, each is a deduction about non-tangible factors that we posit to exist in Nature. Although these are in-tangibles (such as mass or force) they can nonetheless be assessed by infer-ence from sensory data. Sometimes these inferences require use of special instruments to provide evidence. For example, we can measure the effects of the force of gravity on an object with a given mass by placing it on an instru-ment called a balance.

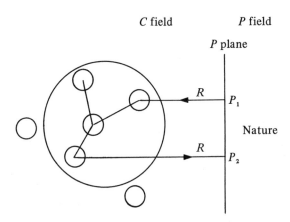

Figure 2.   **The circuit of verification.**   Source: H. Margenau, *The Nature of Physical Reality*, p. 106. Copyright, 1950, by the McGraw-Hill Book Company, Inc. Used by permission of McGraw-Hill Book Company.

The significant characteristic of the small circles within the large one is that they are in some way a direct representation of sensory experience or can be logically derived from sensory experience. This body of logically connected constructs is called reality. Reality is all of our interpretations of sensory experience that can be related to sensory data acquisition. Margenau further specifies how we validate these interpretations of sensory data.

There are two arrows in the figure. One arrow is directed inward from the *P* plane toward a set of interconnected constructs. This represents an interpretation made from sensory experience. The *R* stands for a set of *rules* one uses to make abstract interpretations from sensory experience. There is a second arrow directed outward from the set of constructs toward the *P* plane. This represents a prediction, made from our interpretation of sensory experience, about events we anticipate will occur in the environment. We may predict that an anticipated event will occur in the course of natural events; this is a kind of prediction that can be tested by waiting to see if the event occurs. Or a prediction may be a statement of hypothesis. This is a statement of how variables are interrelated, and it can be tested by determining whether or not the relationship does occur under the experimental conditions we use. In both of these cases some prediction is made based on a set of constructs initially derived from some kind of sensory experience.

Now, if a prediction is shown to be correct — that is, if we actually obtain sensory data again of the kind we anticipate — then our interpretations (set of constructs) are strengthened to the extent allowed by the test. We can say that the set of constructs (theory, model, or other explanatory system) wherefrom we derived the verified prediction has also been verified. If, however, we do not observe the predicted relationship, then we have not verified the set of constructs, and some reorganization of our interpretation of experience is required. For example, a part of a theory on electron scattering by heavy atoms may specify certain angles of scattering of the electrons after bombarding a thin sheet of metal foil. If these scattering angles are not observed, then we can modify our interpretations of electron interaction with heavy atoms to better account for the data that were obtained.

The process of making interpretations from sensory experience (making a logically connected set of constructs), then drawing out a prediction to be tested, and finally assessing the outcome of the test of the prediction to determine the validity of our set of constructs is called the *circuit of verification*. It is a circuit because the object of our prediction is a natural event that we expect to occur based on data gathered from prior experiences. When we test the prediction we receive additional sensory data that tend to confirm or in part contradict the prediction. When the circuit of verification is complete, our prediction has been tested and interpreted to be correct and therefore the set of constructs that yielded the prediction is further strengthened as a part of reality.

There are interpretations that fall outside of the realm that Margenau calls physical reality. Some constructs in religion and the humanities do not have epistemic connections to sensory experience and therefore are not susceptible to verification. The construct of deity, in so far as it is posited to exist aside from direct sensory experience, is not admitted as part of physical reality. Moreover, since constructs of these kinds cannot be logically linked to direct sensory experience, they are not part of scientific explanations of reality. They belong to an entirely different sphere of thinking. Constructs that cannot be linked logically to immediate sensory experience are called insular constructs. These are shown in Figure 2 as small circles outside of the large circle. This view clearly differentiates between scientific ways of thought and those that may be philosophically useful but are not capable of scientific verification.

Margenau's model provides an explanation of science processes that is useful in building curriculum experiences. It provides a rational way of coping with changing interpretations of reality and presents a model of scientific data analysis that can be used to help students understand how scientists operate. Within this perspective, a scientific explanation is always tentative and can be discarded for one that is more useful in making verifiable predictions. In some cases a new explanation is preferred over an older one because the newer explanation is more generalized or elegant than its predecessor.

Suchman has presented a model for the analysis of inquiry that has found use as a curriculum model.[2] It originated as an explanation of the processes children use when confronted with problem situations. The model assumes that the pedagogical environment is open-ended and that there is a minimum amount of teacher intervention. The more teacher intervention, the more the learning environment becomes a didactic experience of information memorization. Suchman suggests a working definition: "Inquiry is the pursuit of meaning." Inquiry is an action of finding relationships between and among separate aspects of one's consciousness. It is further assumed that increments in meaning derived from repeated encounters with the environment are satisfying.

ENCOUNTERS ⟶ ORGANIZER ⟶ MEANING

Figure 3.   **The role of organizers in giving meaning to encounters.**   (After Suchman, 1966).

An *encounter* is an experience with sensory data. The meaningfulness of an encounter depends on the kind of relevant information we already possess. Some kind of *organizer* is required that allows us to select out and pattern certain aspects of an encounter. The application of an organizer to an encounter yields *meaning* (this is pointed up diagrammatically in Figure 3). An organizer is any idea, image, recollection, or other available pattern of

2. J. R. Suchman, "A Model for the Analysis of Inquiry," in *Analyses of Concept Learning*, ed. by H. J. Klausmeier and C. W. Harris (New York: Academic Press, 1966), pp. 177-187.

thought that can add to the meaningfulness of an encounter. As a pupil has more experiences with an object, it assumes more meaning due to the increasing cumulated store of information that can be brought to bear on the object.

An encounter in practical terms is some experience with the environment that appears to be inconsistent with prior experience or provides some novel characteristic that invites the pupil to investigate it. Various *systems* — ways of handling and interpreting data — are used to give meaning to an encounter.

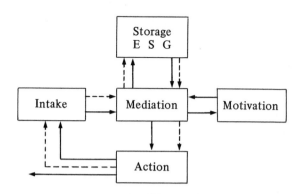

Figure 4.   **The cycle of inquiry.**   Source: J. R. Suchman, "A Model for the Analysis of Inquiry," in J. Klausmeier and C. W. Harris (eds.), *Analyses of Concept Learning* (Academic Press, 1966), p. 179. Used by permission.

A full explanation of inquiry behavior requires the use of mental *storage*. This is a reservoir of encounters, systems, and meanings that one has stored in memory from prior experiences. If a new encounter is to have meaning, recollections from prior experience must be available to provide a context for meaningfulness. The retrieval of organizers from storage must be regulated by a mechanism that governs their selection and application. A selective function is also applied to new experiences to determine which parts of sensory experience will be assimilated. The coordination of memory retrieval and selective information-gathering is called *mediation*. The central function of mediation in regulating intake and the selective recall of organizers is shown in Figure 4. Coordination of *intake* with organizers produces new meanings which are stored for future use and also yields *action* that changes the environment to allow additional intake. The dashed lines represent an inquiry cycle. Action allows the student to manipulate the environment to create new intake. Through mediation the new intake is given meaning (which is stored, thus increasing the total fund of information to be applied in future encounters). The new meanings may suggest new environmental manipulations, thus completing the cycle. *Motivation* influences mediation by sustaining interest in the inquiry process and in turn is strengthened when new input produces a sense of closure (a sense of completeness in the task).

Suchman[3] describes an instance of inquiry learning as follows:

> Consider the example of a child who has witnessed a discrepant event, a demonstration of a blade that behaves in strange ways as it is held over a flame. First it bends downward as it is heated. Then it straightens as it is cooled in a tank of water. The second time heat is applied it bends upward!
>
> The first part of the event is assimilated by the child because he has two available organizers in storage, the concepts of melting and gravity. When combined they provide a satisfactory model to account for the behavior of the blade. As the demonstration continues, the blade is held in a tank of water whereupon it straightens out. It is then inverted and held over the flame again. This time it bends upward, *away from the flame.*
>
> The child is surprised and puzzled. The event is clearly discrepant. He has no single stored encounter, no system, no meaning, in short, no organizer that will enable him to assimilate *in toto* this encounter.
>
> His subsequent behavior can be translated in the terms of the model.
>
> 1. Encounter with blade bending upward.
> 2. Mediation function scans storage for organizer to match encounter.
> 3. No such organizer is available.
> 4. At this point the child usually wants to pick up the blade and examine it more carefully, flex it in his hands, perhaps hold it in the flame again, in short, learn more about the properties of the blade. In terms of the model, he is taking action to generate new encounters, taking in the data and scanning storage for organizers that will make the encounter more meaningful.
> 5. Without success he takes more action and generates more encounters.
>
> In time he will find some organizers that permit him to assimilate at least part of the encounter. He will surely associate heat (a system) with the bending and suspect that the expansion and contraction (two more systems) of the metal are relevant. He might test this theory through various measurements (controlled encounters). In time new data will bring new organizers into play and these will in turn suggest what new encounters are needed.
>
> At all times, the decision as to what operation comes next is made through the mediating function. In other words, the process of inquiry is internally regulated and serves to bring encounters and theories together for matching so that each builds on the other. Whenever a match is made between a theory and an encounter, to a degree the theory is supported and the encounter assimilated.

[3.] *Ibid.*, pp. 183-184.

Inquiry learning requires pupil autonomy in mediation. Any attempt by the teacher to program data input or to instruct the inquirer to use certain organizers tends to convert the process from pure inquiry into some form of externally manipulated learning. Suchman recognizes that both pure inquiry and didactic teaching have merits and that most optimal educational programs will vary somewhere between these two extremes. Free inquiry has the value of requiring the learner to organize and store input for himself and through selection of organizers develop systems for handling new input. In Suchman's view, that input which the student has struggled to organize for himself will have enhanced meaning and retention as compared to didactic learning. As a counter point, I submit that *closure* (satisfactory sense of completion of a problem) is essential if motivation to inquire is to be sustained. An inquiry experience defeats its own purpose if a student cannot find a rational solution in a reasonable amount of time. A too difficult task, or one that has insufficient teacher guidance to allow a rational solution, encourages the student to seek any explanation at all — reasonable or not — to complete the problem and escape the situation. This merely reinforces poor habits of thinking and teaches the student to *avoid* rational and creative data analysis instead of encouraging confidence in finding rational solutions to reasonable tasks.

In summary, contemporary curricula emphasize student autonomy in laboratory experiences. Minimal teacher intervention is sometimes recommended to allow students to seek explanations of experience for themselves. This autonomous function is believed to yield more stable memory of acquired information and to enhance its meaningfulness. The student, moreover, presumably comes to appreciate the tentative nature of scientific explanations and to better understand why scientific explanations sometimes undergo radical revision when new evidence demands it.

It is clear, however, that excessive use of any particular methodology can become boring. Open-ended laboratory tasks that are too difficult for students to complete rationally can reinforce poor habits of thinking and possibly destroy students' interest in scientific experiences. We need flexible approaches to laboratory instruction that will allow the teacher to match learning experiences to a student's level of ability and cognitive style. This is the topic discussed in the next two chapters.

Three

A MODEL FOR
A NEW PERSPECTIVE

The teaching of science is a unique field. The science teacher must be a truly remarkable combination of psychologist, scientist, and philosopher. One must be sufficiently aware of human behavior to assess the current stage of a student's intellectual development, to appreciate the range and specificity of the student's interests, and — with due recognition of the variability in human response — to attempt to create an appropriate expectation of scholarly growth that will motivate the student to learn. Not all students are equally scholarly, creative, highly motivated, or aesthetically sensitive; not all can appreciate fully the value that science instruction can make in their lives. Certainly only a small proportion of the students in our public schools will choose or be able to go on to professional careers in science. Each category of individuals, however, in so far as we can conveniently group people on psychological variables, will benefit from some exposure to the diverse content of science. The student who is bound for a non-science career may never develop the kind of creativity and precision in particular ways of thought that are characteristic of the scientist. Accomplishments in a science class will be varied and relative to students' capabilities. We would hope that whatever experiences we plan would allow for individual directions of interest while aiming in all cases at reasonable growth in logical thinking, rational use of the environment, and appreciation for the organized patterns that science discovers in the environment. The earnest student of science, or one who is an incipient science scholar, needs to have the advantage of a thoroughgoing scientific preparation

17

commensurate with intellectual potential and present state of mental maturation. Such a student will be able to (and deserves to) surge ahead in abstract ways of thought and creativity that will allow early growth in the ways of thinking that are necessary if a mature scientific perspective is to develop. Strong foundations of theoretical thought and precision in the use of evidence at an early age are obvious advantages for one who intends to pursue science as a career. The teacher can certainly benefit by an appreciation of the psychological variables that potentiate as well as limit the kinds of things a student can optimally achieve in a science learning laboratory. Teachers need a thorough understanding of various fields of science, within limits reasonably governed by a specialty that is the main thrust of their teaching. Science is more than organized knowledge, although that is an obvious component that we cannot overlook or reject. Science is a way of thought and a system of methodologies that are given to the task of interpreting natural phenomena. This also is a kind of knowledge, which teachers of science should possess and transmit to their students. It is a knowledge very frequently contributed by philosophers, as cited in Chapter Two. But we need to remember the lessons of Chapter One as well. Understanding science as a human enterprise, we take into consideration the various unique contributions of human intellect to its development. And in science teaching, by the same token, we must remember that students vary in their capability to appreciate and acquire scientific ways of thought.

A teacher in the science laboratory needs a systematic, yet flexible, way of thinking about various kinds of scientific processes that can be matched to students' capabilities. Many of the philosophical analyses available to the teacher simply do not provide a sufficiently graded analysis of science inquiry processes to allow the teacher to make informed judgments about the kind of activity most appropriate for a student or group of students at a given level of scholastic maturity. Some philosophical models seek to explain the most sophisticated kinds of scientific analysis and, in so doing, reach far beyond the kind of activities that many average or near-average students will be able to comprehend or emulate.

It is for this reason that we need to combine psychology with philosophy. We want to generate a model of science processes that combines various levels of scientific inquiry with suitable psychological variables of student ability to yield a flexible view of what a student may be expected to accomplish at any particular point in his development. The model to be presented in this chapter will provide the teacher with a set of science processes that are graded in complexity and abstraction. These processes moreover are combined with psychological and social moral dimensions of science that allow us to present the aesthetic and ethical components of science as a human endeavor.

It is now widely recognized that there is not, simply, a methodology of science. There are many kinds of logical and manipulative skills that a scientist uses in searching out nature. Moreover, there is no single sequence of activities that characterizes a "successful" and "standard" application of scientific method. The kinds of thought processes used and the sequence in which they are used vary considerably from one scientist to another and indeed from one time to another as an individual scientist pursues various scientific problems. There is always that elusive variable of creativity and insight that seems to defy categorization or explanation yet is so central to truly innovative scientific discovery.

Here, therefore, we will not be attempting to describe "a scientific method." Rather, the model to be presented contains a classification of various kinds of science processes that can be employed. They are treated in a certain order — from the more immediate kinds of mechanisms used in gathering sensory data to the more abstract ways of thought characteristic of open inquiry and theory-building. This is not meant to be an invariant sequential model. There are no necessary or prescribed sequences of activities that a student must follow in order to realize adequate or remarkable success in laboratory study. Rather, the model is meant to provide a systematic way of thinking about the kinds of mental activities available to students during science learning and thereby to facilitate the teacher's systematic and informed selective use of these categories in aiding student learning. Since the categories are somewhat graded, from more immediate sensory-data-processing categories toward more complex and abstract ones, some guidance is provided for thinking about matching science thought processes to the intellectual capability of the student. Moreover, since the model provides for flexibility in the sequencing of mental processes, the teacher can help students assess what particular process is most appropriate at any point in a laboratory experience.

With respect to "scientific method," then, the teacher needs to keep this essential principle in mind: *Each student is unique as to his present mental ability, scholastic maturity, and level of interest; and in a given laboratory task there are usually several possible approaches to solution of a problem. Therefore, as a teacher, think flexibly about the kinds of thought processes you can mobilize in the student's mind at a given instant in the laboratory. Help the student who is bemused or confused by the task to identify a next step that is reasonable for his ability and appropriate to the objective of the task.*

Among the various techniques of helping a student to use a suitable method are *prescription* (stating a particular method to be applied), *prescribed selection* (giving two or more possible methods and allowing the student to choose an appropriate one), *induced selection* (indirect questioning that arouses a student's awareness of a method), and *free selection* (creating a learning situation so organized as to facilitate the likely spontaneous discovery of a method by the students).

In this chapter the model is described and the psychological significance of its components is set forth. Chapter Four presents discussion of some methods a teacher can use in applying the model. The four techniques generally defined above will be more particularly developed there.

Dimensions of the Model

There are three dimensions in the model: (1) psychological orientation of an individual toward the natural and social environments; (2) psychological orientation of an individual toward scientific inquiry processes; and (3) a set of scientific inquiry processes. The dimensions will be defined here; their relationships are represented in the three-dimensional diagram shown as Figure 5 (page 22).

Dimension 1.   The psychological orientation of an individual toward his environment is the set of attitudes, beliefs, and values that a person holds about the environment. This dimension is divided into two categories: (1) orientation toward the natural environment and (2) orientation toward the social environment.

One's orientation toward the *natural environment* is the set of attitudes, beliefs, and values that one holds about the non-human environment. For example, one may assume that the natural environment is chaotic or that it is susceptible to being ordered. One may assume that the natural environment is complementary to human technological advance or that it is to be exploited for human advance. The substance of this category is how one perceives the environment as a phenomenon and its relation to his own purposes.

Orientation toward the *social environment* is one's philosophy about the purposes of society and the individual's role vis-à-vis those purposes. For example, a person can perceive society as subservient to individual personal goals or as a cooperative group where the individual adjusts his goals to be consonant with the goals of society at large. Society can be seen as existing to advance its purposes without regard to their cost to the natural environment or it can be perceived as an organization of individuals who manage the natural environment with prudence and wisdom.

Dimension 2.   Orientation of an individual toward scientific processes is the set of psychological variables that determine how a person uses scientific processes. This dimension is divided into four categories: (1) sensation, (2) predisposition, (3) recognition, and (4) operation. The categories are deliberately ordered in a sequence beginning with those psychological variables related to immediate sensory data acquisition and extending toward those variables related to active cognitive application of scientific processes. These categories are defined in the paragraphs that follow.

Sensation is immediate sensory data acquisition through mediation of sensory modalities. Here included are: (a) the preferred kind of sensory modality, i.e., a preference to use one of the five senses in gathering sensory data; (b) the intensity of sensory data acquisition, i.e., the sensitivity exhibited in using a sensory modality; and (c) the duration of use of a sensory modality. Sensation is the most immediate form of data-processing. It is the primary step in selective gathering of stimuli from the environment.

Predisposition is a tendency to behave in a certain way. This is a non-verbalized inclination toward a set of behaviors. It is a tacit psychological state of readiness and a confidence sufficient to sustain activity. This category is akin to what laymen call intuitive understanding. For example, a person may be aware that a certain pattern of plant growth is present in a field observed from a hill but cannot express the pattern in words; at this level of perception the awareness is non-verbal and hence is labeled a predisposition. Another example: an inclination to assume that causal relations exist in experience and to seek instances of them is labeled predisposition. Predisposition is one of the most fundamental psychological variables in this dimension, since it is the requisite primary step toward higher cognitive processing of data. Unless one tacitly accepts the possibility of understanding certain things through experience and unless the necessary confidence is present to sustain such activity, there is little probability that one will perform higher levels of symbolic rational data-processing.

Recognition is a verbalized awareness of sensory experience and of scientific processes that can be used in obtaining and processing sensory data. Whereas predisposition is a tacit, non-verbalized state, recognition is a verbalized awareness of data processing. It includes the ability to explain scientific processes but not the ability to actually perform them.

Operation is the ability to use scientific processes. This represents the highest level of cognitive activity among the categories in this dimension.

Dimension 3. The science processes dimension contains seven categories. These include activities used by scientists in both the physical and biological sciences. The activities are ordered in sequence beginning with those related to immediate sensory data acquisition and progressing toward those that represent higher-order data processing. The categories of activities in this dimension are (1) observation, (2) integration, (3) interpretation, (4) prediction, (5) evaluation, (6) reconstruction, and (7) construction.

Observation is a process of acquiring immediate sensory data. Although this is the most fundamental category among those in the science processes dimension, it is very complex psychologically. It includes the kind and range of stimuli to which one attends, the boundaries that one places on what is to

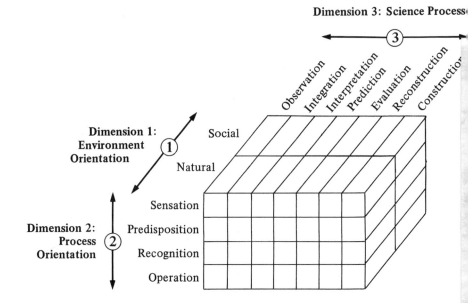

Figure 5.   **Three-dimensional model for a new perspective of science processes.**

be observed in gathering sensory data, and the kind of organization that one immediately identifies in sensory experience. These attention factors are regulated by complex psychological variables. Among these variables are (*a*) prior experience that determines the potency and valence of the stimuli, and (*b*) previously acquired schemas (organized ways of viewing the world) that guide selective attending and determine the range of stimuli attended to and the immediate organization one imposes on sensory data. These factors are set forth in much greater detail in a following section devoted to more thorough discussion of Dimension 3 categories.

Integration is a process of relating newly acquired data to prior-gained information. This is a transition category between observation and interpretation. Integration is a process of associating immediate experience with stored information without performing elaborate logical analyses about the kinds of relationships that exist. Integration includes first-order associations such as identifying an instance of something previously encountered. Higher levels of integration include classification of data in previously formed categories. To identify a spider as belonging to a particular genus is an act of integration. To see a larger-than-normal specimen of an animal and thus to realize that a given class of living things contains a wider range of sizes than previously understood is a high-level act of integration that approximates the next higher process category: interpretation.

Interpretation is a process of deducing relations among immediate sensory data or between newly acquired data and prior knowledge. The relations are of a higher order than simple associative relations and assignments of data to given classification categories. Interpretation includes the following kinds of processes:

 a. Invention of new classification systems or categories for ordering sensory data.

 b. Application of mathematical systems such as statistical and graphical methods of data analysis.

 c. Synthesis of temporal or spatial relations among sensory data (this includes "explanations" of observed forms and functions).

 d. Statements of correlation or cause and effect among variables that are not merely memorized data relations (this comes very close to the next higher process category: prediction).

Prediction is a statement of events that are expected to occur or a statement of cause and effect that can be subjected to a test of verification. At the lowest level, this category includes simple extrapolation, i.e., an extension of a trend beyond acquired data. At the highest level of prediction stands a hypothetical statement containing clearly defined independent and dependent variables whose causal relations are set forth with due attention to contributory variables that are to be controlled or accounted for during a test of the hyphothesis.

Evaluation is a judgment of the accuracy of a prediction or the quality of sensory experience in relation to a criterion. The criterion can be a standard of quality, a value statement of what is good or acceptable, or an objective that is to be accomplished.

Reconstruction is a process of reorganizing or otherwise refining a theoretical explanation or model in the light of newly acquired data. This category presupposes the existence of a set of ordered constructs and of newly acquired data that present evidence requiring reorganization of the constructs to better accommodate the data. It is presumed that part of the original theory will remain constant and that the changes made do not represent a radical revision to the extent that the theory would be classified as overthrown.

Construction is the synthesis of a new theoretical statement or model. It is such a dramatic departure from prior explanations, or includes such novel interpretations of sensory experience, that it can be considered a new statement. Construction is the production of a set of orderly relations among constructs and as such implies an explanation that is more complex than a new classification scheme as categorized in the discussion of interpretation.

The categories in the three dimensions can be combined in various ways to produce ternary groups containing one category from each dimension.

These combinations provide flexible ways of thinking about student logical thought processes that can be used in laboratory teaching.

The foregoing preliminary definition of the categories and their assembly into the cubic model in Figure 5 will hopefully serve as an organizer for the more particular discussion that is to follow: (1) The categories of each dimension will be discussed from the perspective of psychological theory, and their implications for teaching will be set forth: (2) The categories will be integrated in binary and ternary combinations to specifically illustrate ways of generating learning strategies for students in a laboratory experience.

I believe this presentation of successive approximations to a complete description of the model will allow at the conclusion a better understanding of its use than would be possible without such a distributed discussion of its components and their relationships. Meanwhile, however, the reader may find it helpful to scan Chart 4, which provides in concise form a complete outline of the model's categories and subcategories (page 102).

Categories in Dimension 1:
Environment Orientations

Each of us views the world as it were from an internal perspective, much as one stands within his house and views the outside world through a window. We set outselves apart from the world as individual beings. Each of us has a particular self-identity which, depending on our social maturity and objectivity, is more or less clearly defined as to interests, abilities, attitudes, life style, patterns of work and amusement, and many other conscious categories that allow us to represent ourselves as rational beings. It is particularly characteristic of Western thought to posit man as a separated viewer of the world; and thus we bring a certain set of mental perspectives to bear on what we perceive in experience. I will not argue here the relative merits of such a separatist view of the world as compared with a synthesist view that sees man as more fully immersed in, if not one with, the physical universe. Assuming that to a greater or lesser degree we accept the separatist view — that we are beings who receive stimuli from the world and make interpretations of these — then it is of interest to ask what are the perspectives that we bring to bear upon the reception and analysis of sensory experience.

For convenience I have selected two aspects of the environment that are pertinent to science laboratory teaching. These are the individual's orientations toward the natural and social environments. There are obviously many subcategories that one could identify. Only those most pertinent to the objective of this book will be presented. I will begin with those that are more related to affective psychological variables and proceed toward those that are more cognitively oriented.

Natural Environment Orientation

Aesthetic Perception.   Science is at least partially an aesthetic experience that is mediated by careful intellectual investigation of the environment. Laboratory science instruction should provide opportunity for students to develop a new view of the world — one that includes orderly ways of appreciating organization in Nature and also includes tolerance for the variability in form and temporal relations that the environment presents. Science learning without question is an intensive endeavor requiring a good proportion of intellectural devotion in its attainment. But it can also be an endeavor producing deep satisfaction through experiences that evoke wonder, and even a joyous sense of marvel, at qualities and patterns discovered in Nature. A wise teacher makes reasonable provision for these more emotive components amidst the more traditional, purely intellectual, pursuits of science.

Some concepts that may be involved in developing aesthetic response in science learning will be discussed here: (1) symmetry, (2) complementarity, (3) balance, (4) periodicity, (5) concentricity, (6) reticularity, (7) pattern-unit relations. These concepts can be used as themes in studying a wide range of phenomena.

Symmetry is correspondence in size, form, and arrangement of parts on opposite sides of a plane, line, or point; any form that can be divided into two or more mirror-image components is symmetric. Plane symmetries are those forms that can be divided into two or more corresponding parts on opposite side of a plane. Rotational symmetries are those forms that can be rotated through angles less than 360° and brought into correspondence with the initial state. Examples of systems exhibiting plane symmetry are the arrangement of p-orbitals in an atom, the bilateral symmetry of animals, the opposite leaf arrangement in plants, some gem-disubstituted molecules in chemistry such as gem-dichloroethane, pairs of optical isomers, and other forms that can be divided into halves of mirror-image relation. Examples of rotational symmetries are the spiral arrangement of some protein molecules (α-helix), the compositae flower head (this can also be plane-symmetric), whorled and spiral leaf arrangements in plants, spiral nebulae. It is important to recognize that *spirals* of various kinds are a special instance of rotational symmetry. Spiral forms occur so frequently in biology as to warrant special effort to allow students opportunities to identify them. Moreover we must recognize immediately that their presence usually aids survival by providing greater strength or more equal and evenly spaced distribution of parts. Ring thickenings on the walls of insect trachea and spiral and ring thickenings on plant tracheids give strength against collapse. A spiral leaf arrangement in plants distributes the leaves in a way that optimizes light collection. In physical systems, spirals of appropriate elasticity (such as coil springs) can be used to store potential energy.

Complementarity is a relationship of two or more forms such that each form supplies something lacking in the other (a template-product or lock-and-key relationship). "Form" is used broadly here to mean a solid form or a function found in Nature. This usage is very general indeed and some further explanation is warranted. We usually consider form as being the spatial arrangement of matter. However, the term can be used much more generally to mean any organized and reproducible occurrence in the environment. Thus, in this sense "form" means any identifiable organized component of experience either material or abstractly conceived. A process, an interaction pattern among animals, a developmental sequence, and a concept such as dominance and dependence in behavior are considered here as forms that can exhibit complementarity. Examples of complementarity are the base-pairing between strands of DNA and messenger RNA, symplastic growth of cells where each accommodates to the shape of its neighbor cell, overlapping environmental niches that do not lead to competitive destruction of the inhabiting species, enzyme-substrate complex, enzyme-coenzyme association, commensal relations of organisms as in the algal-fungal relation in lichens, ionic bonding where one atom effectively loses an electron and one gains an electron, and coupled half-cell reactions in electrochemistry.

Balance is a state of equilibrium or equipoise; equal distribution of weight, amount, and so forth. Examples include equal distribution of forces on a line around a point, a pattern of tree growth with the same number of limbs on either side of a plane, or a state of an ecosystem when nutrient supply is sufficient for nutrient demand.

Periodicity is a regularly repetitive pattern in space or time. Spatial periodicity is exemplified by regularly spaced objects in space, such as body segmentation in a worm, insect wing scale arrangement, and stratified horizons in geological formation. Temporal periodicity is a regularly spaced series of events in time. Included here are light and dark cycles, plant and animal life cycles, and various rhythms in living things. Examples from physics include various forms of wave motion, planetary movement, and oscillation.

Concentricity is a condition characterizing a set of circles, spheres, cylinders, or cones that have a common center, with larger forms enclosing smaller forms. Examples of concentric circles are rings observed in the cut surface of an onion bulb, annual rings in trees observed in a transverse section, starch grains observed in plant cells, circumferential lamellae observed in animal bone transverse section, Bohr's model of the atom, the planetary orbits of the solar system, layering in cross-section of rocks as in certain geodes, wave formation on a liquid surface caused by a pulsating point source. Concentric spheres, cylinders, and cones are exhibited by solid objects and usually must be inferred from cross-sections through the object. For example, onion bulbs are actually approximately concentric spheres of scales; a tree trunk contains annual growth layers that approximate concentric cones; the body plan of some coelenterates and segmented worms is a set of concentric cylinders. Concentricity is a special form of spatial periodicity, since it presumes a regularly spaced set of inclusive and subsumptive repetitive forms. *Reticula (networks).* A reticulum is a set of interconnected points lying on straight lines that form a net-like arrangement. Reticula are found in many biological systems such as the growth of *Hydrodictyon reticulatum,* leaf venation, blood capillary networks, plant tissue cell wall organization, and fungal mycelial growth. Electrical circuits can also be considered reticula. *Pattern-unit relations.* "Pattern" is used here to mean any organizing principle whereby instances of experience can be ordered. A "unit" is any particular observation that can be subsumed by an organizing principle. A pattern-unit relation is posited when one finds instances that can be subsumed. As an example, the concept of "succession" in ecology is a pattern, since it can be used to identify various instances of plant growth sequences. The concept of "field" in physics is a pattern, since it can be used to explain various kinds of magnetic, gravitational, and electrical phenomena. Pattern identification is one of our most fundamental cognitive processes. Gestalt psychology has clearly established that humans interpret spatially arranged points as belonging to organized groups. The way we group things depends on various spatial and logical factors, but among these are the spatial proximity of objects, and the similarities of objects, in form or other properties, that allow us to deduce organizing principles to explain their relationships. In some cases we use temporal or functional characteristics of phenomena to generate explanations of their relationships. There are various abstract qualities of phenomena that we sometimes use to interrelate them and explain their organization in space or time. One instance of a temporal abstraction is the previously cited concept of succession.

Organizational Perception: Order-Disorder. Individuals' perspectives vary with respect to the presence of order in the environment and the ability to tolerate variability within observed natural environment data.

Each person brings a unique mental set about the environment. In some cases life experience has been so chaotic, or so little attention has been given to ordered events, that the person does not have an understanding of orderliness as a concept. An obvious basic condition for scientific inquiry is at least a readiness to perceive organization in the environment. However, one must also have a tolerance for variability. Many phenomena can be ordered to a degree, but there exists a residuum of disorder that seems to be inherent in natural events. An educated person is able to comprehend the degree of disorder that persists and make some assessment of its amount. This can be a subjective assessment of "greater than" or "less than" that observed in some other system, or it can be the use of statistics to express the degree of uncertainty quantitatively, as by an "error of measurement."

Social Environment Orientation

Value Judgments. Value systems are statements of belief about man's obligations, as a social being, to sustain social organization, plan social progress with consideration for environmental impact, control disease, and alleviate human need.

I will not attempt to set forth a particular value system in this book. Each social community will no doubt have some common value beliefs that students can bring to bear on discussions about the impact of scientific discoveries on society. Moreover, there are some recognized values of precision and honesty in scientific research which can be brought to bear on issues of quality in scientific inquiry. Although technological applications of scientific findings are not usually considered part of natual science, it is clear that some attention should be given to the impact of scientific research on society and the natural environment. The Social Environment category in Dimension 1 provides for such consideration.

As an aid to teachers' use of this category, I present some conceptual information on value systems. This is intended to give some general guidelines for assessing students' rationality in using value systems. For purposes of our discussion, we will need to describe value systems sufficiently broadly to allow application both to human behavior generally and also specifically to scientific activities. We divide value systems into two kinds: (1) those systems of belief that posit what is good and/or desirable human behavior, and (2) those systems that posit what are cherished and/or valuable artifacts of man's activity. The first category has to do with quality of human conduct, the second with the quality of human artifacts – the things people produce. It is obviously useful to make a distinction between the quality of a person's conduct that contributes significantly to the health and cohesiveness of society and those material things and intellectual products which potentially are no less important to society but are not directly a part of human personality structure. As a simple example of these two ways of understanding persons, we might say of our neighbor: "He's a fine law-abiding and honest man, but he certainly cannot grow a fine crop of tomatoes."

Those systems of human conduct that posit what is ideal and reasonable behavior are called systems of ethics. These are the conceptual statements of what is desirable in human conduct. Statements about the specific kinds of activities that an individual should or should not engage in are called statements of human morality. A system of ethics, then, is a kind of general guide as to what is desirable in human behavior, whereas a statement of morality is a specific explanation of how a person should conduct himself.

As a teacher, you can do much to help students develop a logically consistent and well-thought-out system of ethics to guide behavior. One of the basic requirements of such a system is that it be logically consistent. That is, if a certain condition is held to be desirable today, then tomorrow under similar circumstances it should be desirable. Consider this purely hypothetical example of logical inconsistency: a student may state that it was acceptable for Dr. X to falsify some data, since he subsequently earned a Nobel Peace Prize, but not acceptable for Dr. Y, who worked in a small industrial laboratory of little known accomplishment. A sensitive teacher can be quick to follow up on such a statement with some discussion of why honesty is so critical to good scientific research regardless of the social esteem a person may achieve. The ethical requirements of honesty are built on the assumption that science is a community endeavor wherein the contributions of one person are used by others to further our understanding of the natural environment. If people are not honest, then the systematic and orderly cumulative understanding of the environment will fail. For this reason, the requirement of honesty is equally placed on all scientists.

In addition to logical consistency in statements of ethics, there is the requirement of logical accuracy in drawing deductions about what is desirable in a given situation. Consider, again, the matter of honesty in research. A student may accept the assumption that all scientists are to be equally responsible in honest reporting of their research, but then may fail to logically relate industrial scientists to the broader class of scientists in general. A statement such as "It was all right for scientist A to falsify his records because he worked for a company that makes cosmetics" is obviously false in light of the original assumption that *all* scientists are to be equally responsible for honest data reporting.

Although I have promised not to set forth a particular value system in this book, there is one point that needs to be kept in mind. In many societies, particularly those that endure, ethical systems cohere around the ideal that the principles of human conduct that are most desirable are those that place human welfare of all group members above the particular satisfactions of an individual or small group of individuals. To the extent that students utter value statements antithetical to this basic belief, some discussion should be devoted to the importance of making our individual goals consonant with the larger altruistic goals of society.

With respect to value judgments about the artifacts of society, which of course include those of science, the same basic criterion is applicable: to what extent are the products of society in the general best interest of society? Scientific endeavors are of course susceptible to this evaluation as are other kinds of artifacts. This raises an issue of comparative importance between scientific freedom to inquire at will and the effects of such inquiry on the social best interest. This is a topic worthy of discussion in the science classroom, particularly if a current event suggests a possible contradiction between social best interest and the desire of scientists to freely inquire. We have seen such issues strongly debated in public forums. For example, to what extent should massive nuclear testing be allowed when the radiation emission may be harmful to health? In medical research there are questions to be answered about the propriety of using human subjects in research and about what limitations should be imposed as to seeking their consent and making them fully aware of the consequences.

Categories in Dimension 2:
Process Orientations

The categories in Dimension 2 pertain to the way a student approaches and implements scientific problem solving. It contains the psychological factors that limit or enhance use of scientific inquiry processes as categorized in Dimension 3.

Sensation

Sensory data are obtained from the environment by sensory receptors. Photoreceptors (the eyes) gather light-mediated data, mechanoreceptors (ears) gather auditory data, tangoreceptors (fingertip organs of touch) gather tactile data, and chemoreceptors gather olfactory and gustatory data (smell and taste). Each of these avenues of sensory data reception is called a modality. People vary in the kinds of modalities they use most keenly. Some people tend toward visual dominance in reception while others tend toward auditory dominance. Moreover, through experience we learn to use particular modalities over other ones in a given environmental setting. In most class-rooms, visual and auditory modalities are favored because of the talking and visual presentation that characteristically occur there. In the natural environment a person may use tactile and olfactory data-gathering to a much greater degree than in a usual classroom setting. The laboratory is a useful place to enhance wider use of sensory modalities and to develop an appreciation for their appropriate and precise application.

In addition to the kinds of modalities people prefer to use, they also bring to bear on tasks mental sets limiting the kinds of data to which they are sensitive. For example, a student may be very keen about color perception but generally fail to notice slight gradation in albedo (reflectance properties of objects). In visual data-gathering, some students may tend to focus on one or a few objects in the field of vision and neglect to consider other objects more peripheral. They exhibit channel-focused vision rather than collecting data widely when this is appropriately useful. Some students may tend to specialize in visual data-gathering and neglect auditory cues. In the field environment, for example, they may tend to identify animals by sight alone rather than combining sight with sound. Sensation is categorized here as a fundamental or immediate level of psychological processing of data. It is intended to help the teacher identify idiosyncratic sensory behaviors of students and help them to better apply sensory data-gathering skills.

Predisposition

Much of our psychological processing of sensory data is mediated by verbal representation. We have language which we use to represent experience and to allow symbolic transformations to yield new logical deductions. However, not all of our processing of experience is verbal; or at least some of it is not available to conscious explanation of what we are doing when we react to a situation. There are some tacit or non-verbal ways that we use to approach and analyze an experience. These tendencies to behave in a certain way at a non-verbal level are called here predisposition. Predisposition represents the kind of attitude and inclination to behave in a certain way that we bring to bear on a situation.

At the most fundamental level, predisposition is an inclination akin to confidence. It is a sense of self-assurance that provides sufficient drive to allow an individual to begin a task. It is a sufficient sustaining motivation to allow one to continue in a task under reasonable conditions of accomplishment. The laboratory environment provides a context that can facilitate enhanced confidence in approaching a problem. If students are given reasonable tasks commensurate with their intellectual abilities and scholastic preparation, much progress can be made toward building a predisposition of confidence toward problem solving. It is equally obvious that repeated defeating situations can build an equally opposite predisposition of timorousness.

At a more advanced level within this admittedly fundamental category is an awareness of the availability of certain ways of thinking and certain methodologies of investigation. In layman's terminology, the student has an intuitive sense that something exists or can be used in a situation. For example, there can be an awareness that orderliness can be found in Nature. There may also be an inclination to seek patterns and other orderly interpretations of experience. Since this is sometimes a tacit understanding, the teacher can detect its presence only by (1) observing the attitude that a student brings to a task (does he fumble around excessively? does he show signs of timorousness?) and (2) listening carefully to the kinds of ultimate explanations the student gives for experience. Chaotic or highly non-rational explanations may indicate that the student does not anticipate finding an orderly explanation of experience and hence does not produce one.

Recognition

Our verbalized store of information is available for explicit presentation. We can talk to others about what we know and what we can do. Recognition represents the verbalized awareness or definition of parts of experience. A student who recognizes that a certain method of scientific investigation can be used is able to explain the method. Here is required only the ability to define and describe parts of experience. The student is expected not actually to demonstrate an ability to reproduce a part of experience by manipulation, mentally or physically, but merely to state its characteristics. This category is a verbalized awareness of natural phenomena and methods of investigating natural events.

Operation

There is a clear difference between being able to describe something and actually being able to create or perform that which one describes. Operation is the most sophisticated category in Dimension 2. It represents performance level capability with respect to use of science processes. A student is expected to be able to actually perform a particular kind of skill in analysis. To apply a method of investigation efficiently implies more than being able to call it forth at request. There is a question of the degree of sophistication in its use. The degree of sophistication of operation is determined by the extent to which an individual can (1) explain the purposes of an operation, (2) differentiate when an operation is appropriate or inappropriate, and (3) modify an operation when the situation demands it. The teacher should be alert to note when and how a student applies a certain operation. If a student consistently uses a method under inappropriate conditions or seems so inflexible that he cannot modify its use when new conditions require it, then you should be aware that the student has operational capacity but not with sophistication. Some help is indicated to increase the student's understanding of when a method is appropriate and when it is inappropriate. Moreover, assistance should be given in helping a student modify his application of a method to make it better suited to the conditions.

Categories in Dimension 3:
Science Processes

A consideration of various science inquiry processes is given here. Where appropriate, more intensive discussion is devoted to the psychological variables associated with a category. This dimension encompasses some of the more traditionally recognized processes of science inquiry. A thorough understanding of this dimension will facilitate comprehension of the two- and three-dimensional Category Combinations discussed in a later section.

Observation

There appears to be little agreement as to the meaning of "observation" in science. Some writers limit its meaning to looking at something and describing what is there. Others make a much broader definition to include not only data gathering and description but also interpretations of what is being observed. In this book I define observation as a fundamental science process of gathering immediate sensory data and organizing it in very rudimentary ways. To observe is to use the senses in the most elementary way of collecting sensory data and detecting fundamental organizational schemes in the percepts.

When one observes something by visual examination, auditory attention, and the like, it may seem that this is a very simple process as well as being a very fundamental cognitive function. It is indeed fundamental to higher levels of intellectual information processing, but many psychologists recognize that the psychology of observation is rather complex. To look at something, for example, is not a simple matter of pointing one's eyes toward an object as though they were an impassive camera ready to record whatever visual pattern is focused on the retina. We look at our environment in a selective way. There are some things which regularly attract our attention, whereas other things may be systematically overlooked. Once we have actually focused our eyes on an object there are many complicated processes of cognition (mental activity) that give form and organization to what is observed. Given the very complex nature of observation, some careful attention is given here to the psychology of this process. I will discuss three aspects of observation: (1) attention, (2) perception, and (3) cognition.

Attention.   The first step in sensory data-gathering is to orient our sense receptors to receive stimulation. The act of selectively applying a sensory receptor to a stimulus source is called *attention*. If a bird suddenly flew within my field of vision, I most likely would turn my eyes toward its path of flight and focus upon it. What is it that determines my focus of attention? How and why do we selectively attend to one thing and not another?

Stimulus novelty is one factor determining attention. If something new appears within our sensory field, we usually attend to it. A sudden unanticipated movement, or an unexpected sound, usually draws our attention. Neurophysiological analyses of these responses suggest that our nervous system is peculiarly organized so that novel stimuli are detected by certain cerebral cortex neurons which send impulses to the lower brain reticular formation. The reticular formation excites vegetative and cognitive arousal and motivates the orientation of the appropriate receptors toward the novel stimulus source. We are "wired" as it were by nature to actively seek after a novel stimulus. Novelty — in moderate amount — stimulates attending.

Yet we all recognize that novelty alone cannot account for all of our attending. When we see a familiar face in the crowd, we attend to it. When a favorite strain of music begins we listen to it, sometimes very attentively, although we may have heard it many times before. There are, however, some objects of great familiarity that we ignore, and some indeed that because of our prior experience with them arouse a definite rejection of our attention.

What governs these selective attention processes? This is a very fundamental question for a science laboratory teacher. It lies at the very root of successful observation. Much of the success of a laboratory experience is determined by the readiness of the students to attend to the task at hand and make relevant and appropriate decisions about what aspects will be examined in detail. We therefore give considerable space to discussion of selective attending.

Prior experience determines to a great degree what we attend to — and also indeed how we organize the data we receive. For the moment we will discuss it only in connection with selective attention. How we organize our sensory data will be dealt with when we discuss *perception* as an aspect of observation.

Our storage of information from successive experiences cumulates over time. In some cases this information cumulates as specific verbalized representations of experience. We store rather clearly some experiences and can generate rather detailed narrative to recreate the experience as well as we can remember it. We must realize of course that even these apparently detailed recollections of experience are not invariant but undergo transformations in time so that what we recall on one day about an experience can be quite different from what we recall about it on another day.

In addition to these apparently detailed recollections of experience, we also carry more generalized representations of what we have experienced. These global or generalized "views" of experience are called *schemas*. A schema is a representation of prior experiences that (1) governs our selective attention to stimuli, (2) determines the range of stimuli to which we attend, and (3) partially regulates the duration of our attention. A schema is in a vernacular sense a kind of mental map that we develop through experience with the environment. It is an organized representation of experience that contains "landmarks," as it were. These are the aspects of prior experience that have saliency for us. They are the general classes of stimuli that either have positive valence (attract us toward them) or negative valence (repel us from them). In addition to the salient landmarks in our schemas, we also have certain "familiar pathways" that we anticipate in the environment. We tend to expect that things we have seen together temporally or spatially in the past will again appear in this relationship.

A schema is a generalized way of representing kinds of stimuli. It includes (1) the attractiveness of certain kinds of stimuli for us and (2) the patterns of stimulus groups or of isolated stimulus occurrences that we anticipate will recur. Each person has unique schemas, since each person has been exposed to a different total set of experiences. But people of a similar cultural background, of common community origin or small group association, also share some schema aspects in common. We all differ in particular ways, but most of us share some expectations and generalized ways of representing experience in common with our colleagues in a community. To the extent that we have been exposed to similar experiences and have engaged in mutually earnest dialogue about these experiences, we come to develop similar schemas.

To further clarify this concept, I present some verbal descriptions of the kinds of mental constructs that constitute a *schema*. It is not possible to fully describe the functions of a schema in guiding our behavior, since they are largely tacit. They are nonverbal generalized ways of responding to experience, psychological modes of the kind we have called *predisposition* in this book. Nonetheless, it may be helpful to cite those characteristics of schemas that I have metaphorically described as mental maps with landmarks. Since we are concerned with science instruction, I will use examples relevant to that.

Consider the case of a student who is about to examine the various life stages of the common mosquito *(Culex)*. The schema he brings to bear on that experience includes all of the knowledge previously acquired and the approach or avoidance tendencies associated with the insect. Assume the student knows that the mosquito lays its eggs on water, that the young hatch and live in the water for some time, and that the swimmers then change to a form that flies. The flying form, moreover, he full well knows will bite, leaving an obnoxious whelk that itches and disappears after a day or so. He may consequently have a distinct aversion toward the flying form of the insect but little or no emotional reaction to the swimming form. Assume that the student knows nothing more about *Culex* other than this. However, in a broader context, the student's schema is sufficiently undifferentiated that he presumes that all swimming creatures have gills to be used in obtaining oxygen ("breathing"), as in the case of many fishes.

Given these understandings from prior experience, then, the student approaches the task of observing a jar containing various life stages of the mosquito. The goal is to order them in sequence and explain their life activities as deduced from his observations. A secondary goal is to explain why spreading a thin film of oil on the surface of water will kill the mosquito during the swimming stage. Assume that the student has learned that most living things that undergo metamorphosis increase in size and complexity as they mature; thus he should usually be able to deduce the proper sequence of development from egg to larva to pupa and adult stages. A distinct disadvantage is encountered, however, when the student attempts to apply prior understandings to the present observation. Assuming that all swimming creatures have feathery gills to obtain oxygen leads him to describe the antenna of the larval stage as a gill and to overlook the siphon used to obtain air by projecting it above the water surface. The gills of the larval stage, moreover, not being particularly feathery, are misinterpreted as flippers. The student correctly deduces the role of the abdomen of the larval and pupal stages in producing locomotion by an oar-like action.

The student's deductions clearly exhibit an undifferentiated schema; some guidance is required to help him further refine his interpretation of aquatic forms of life. This can be accomplished by telling the student to carefully observe how the aquatic stage of the mosquito approaches the surface of the water. Through careful observation, the role of the siphon can be determined, and consequently the student can also understand why a thin film of surface oil causes death by preventing protrusion of the siphon into the air. Further schema differentiation occurs when the student recognizes that the adult form consists of a larger size female, which does the biting, and a smaller unobtrusive male that does not bite humans.

Consider as another example a student who is about to observe and describe the formation of waves in a ripple tank during a physics lesson. The student's prior experience with waves is limited to surface waves driven by wind and surface waves on a pond initiated by a tossed pebble. He therefore assumes waves are a traveling disturbance that rolls across the surface of the water. When confronted with a standing wave and interference patterns produced by interaction of waves, he is unable to give an adequate explanation for the observations.

The purpose of the foregoing narrative is to make two points: (1) A student's prior experiences with phenomena can facilitate or impede his accurate and rational interpretation of experiences, depending on the extensiveness, accuracy, and degree of differentiation of these schemas. (2) A perceptive teacher attempts to deduce the student's schema by noting his ways of interpreting phenomena, obvious omissions in observation, and clear inconsistencies or errors in interpretation. From these inferences, the teacher can help the student develop a better interpretation of phenomena and learn how to evaluate prior experience as a guide to present analyses. The teacher should serve as a guide to help the student clearly recognize the fine differences among observations and to store them in memory in such a way that these differences will be preserved. Moreover, the teacher can further aid the student by helping him increase the total fund of information stored, thus increasing the scope of his schemas. Chapter Four contains additional information on the use of the concept of schema in planning laboratory learning experiences.

In our discussion of attention thus far, we have noted that a given student may attend to something in his environment either (1) because it is novel or (2) because it is salient for him through prior experience, and (3) if it has some organization that is at least minimally familiar. That these conditions are different for different people is something the teacher needs to keep clearly in mind.

Some students are very compulsive about the kinds of stimuli they will attend to. If a stimulus pattern (any experience with natural phenomena) is very different from their expectations based on prior experience, they may tend to reject it. The stimulus pattern does not match their schema; therefore they do not attend to it. Other students are much more flexible about attending to stimulus patterns that fall outside their expectations. A student who consistently "does not see" a phenomenon may be neither uncooperative nor unmotivated: he may simply have a very narrow mental set that precludes sensing things very different from his expectations. A wise teacher will take time to prepare such students for new experiences by reinforcing each approximation toward more comprehensive pattern acceptance and pointing out the advantages of keeping an open mind about accepting seemingly heterogeneous experiences.

There are at least two phases to attention: (1) preattentive processes and (2) focal attending.[1]   Preattentive processes are generalized ways of attending to phenomena. They are the diffuse attending responses we make before focusing on a given aspect. For example, we may be generally aware of a set of objects that are familiar to us and represent them in terms of their meaning for us based on the schema we have. As a particular instance, when we enter a familiar room, we need not focus our attention on each object to recognize that it is there. The general set of objects is sensed by peripheral vision and collective sound reception that affirms their presence. This generalized attention response allows diffuse recognition of familiar objects in our environment. Focal attending, by contrast, represents an intensive examination of sensory experience to determine particular aspects of it. For example, in focusing our eyes on an object we center an aspect of it on the very sensitive foveal area of the eye to gain as much detailed information as possible. We do not depend on the collective representation from prior experience (schema) to give meaning to the object; rather, we examine in detail the stimulus source to gather as much novel information as possible. Students can benefit from a recognition that focal attending will enhance data reception and allow more particular understanding of a phenomenon than is possible if one relies solely on generalized observation. Focal attending is usually preceded by preattentive processes of gathering diffuse information from the environment.

[1.] U. Neisser, *Cognitive Psychology* (New York: Appleton-Century-Crofts, 1967), p. 86.

Perception. When a person focal-attends visually, he fixates an object on the retina for close inspection and differentiates certain parts of it from their surroundings. Some such sensory process — visual, aural, or tactile — of segregating phenomena and their parts, thereby giving them identity, is the first step in *perception*. Perception is the process of detecting form and physical properties or organization in what we attend to. When a person focal-attends toward an object and detects it to be a cube, he has perceived the object as having a certain form. Perception also includes judgments about size, weight, texture, and color properties of an object. Here also are included spatial relations among objects — their relative distances from one another, side to side and front to back in the sensory field. The awareness of patterns is another important function of perception.

The kinds of organization (if any) that a person senses in an experience are determined partially by prior experience. The schemas already acquired by the individual sometimes bias his perception. One student may examine a mineral crystal and declare that it looks like a grain of salt, whereas another may declare that it looks like a diamond. Perception is not a uniform response for all people. It is particularly important to recognize, in teaching science in the laboratory, that students will bring diverse organizing schemas to bear on perception. Such variability in behavior is to be tolerated and indeed encouraged in so far as constructive and creative products are realized from this diversity. Sometimes, however, it is useful to help the students develop some expectations about a phenomenon before they actually begin observing it. This suggests mobilizing and refining the schemas of the students to help them selectively attend to and efficiently perceive certain aspects of a phenomenon. I will discuss procedures of this kind in Chapter Four.

In general, it is useful to analyze student behavior to determine tendencies toward extremes in perception. Some students may frequently perceive only "units" in the perceptual field; that is, they tend to fragment phenomena into isolated segments. This is indicated by a tendency to cite a list of things (unitary objects or events) that were observed. Other students, in contrast, may see only "patterns" in the perceptual field; they regularly speak about organized wholes. They see interrelationships among events and objects under observation, but do not speak much about the details of any single (unit) object within the whole.

We categorize these two kinds of perception respectively as *unit-dominant* and *pattern-dominant*. A student who consistently organizes experiences at one of these extremes should be encouraged to look for alternative explanations of what he senses.

To further explain the difference between unit-dominant and pattern-dominant perception, I will cite examples of both kinds of behavior. Assume that students are presented with an insect to be observed. A unit-dominant individual might report that he sees "a bug with six legs, a gray body, two wings, a head with two eyes and two feelers." A pattern-dominant individual looking at the same organism may report "a bug whose body has joints all over and is able to walk or fly." Consider another example: students presented with a picture of a spiral galaxy and asked what they see. A unit-dominant individual may report seeing "some bright stars, some lighter stars, some small stars, and some large stars." A pattern-dominant individual might see "a spiral arrangement of stars with curved arms and a very dense mass of stars in the center of the spiral."

The foregoing examples involve visual perception, but it should be clear that the same principles apply to hearing and touch. Some students may hear patterns of sound but not the unit components that make up the sound; others may distinguish the components but not sense the patterns. Similarly, tactile perception may be unit-dominant (sensing only individual parts of a surface, such as the number and kinds of elevations on a surface) or pattern-dominant (sensing the arrangement of elevations on a surface). Although hearing and touch are less frequently employed in data-gathering, their appropriate use should be encouraged.

In some respects the pattern-dominant individual is more sophisticated than the unit-dominant person. The pattern-dominant person exhibits greater perceptual abstraction. However, the pattern-dominant person may be missing important detailed information. Therefore balance in perception is desired so that both unit and pattern perception can be combined during observation.

Unit perception involves the identification of unit properties — properties of individual objects or events, or parts of objects or events — without concern for relationships among the individual phenomena. Unit properties of objects, for example, include color, shape, size, weight, density, and texture.

Pattern perception is the identification of pattern properties — organizations of interrelated units — using two or more of the following processes: *spatial tracking* (scanning a set of spatially connected units); *grouping* (assembly of units into clusters according to common properties or functions, observed or inferred); and *ordering* (assembly of units, or clusters of units, according to a principle of relative size or shape, inferred importance or temporal sequence of functions, and so forth).

Pattern perception as analyzed here may be a new concept to many readers, and therefore some examples of each subcategory are given. *Spatial tracking* is a very primitive kind of organizing activity. It involves the ability to follow a set of interconnected objects in space and to abstract the general configuration of their arrangement. As a particular example, a student who has dissected an earthworm by opening a dorsal incision can then, through spatial tracking, see clear evidence of an alimentary canal and the circulatory system. Spatial tracking in this case may lead to *ordering* — conceptualizing the general sequential organization of each of these systems. Another kind of spatial tracking involves the ability to conceptualize sequential temporal events in terms of spatial forms or properties associated with the observed temporal phenomena. This is exhibited by a student who observes the motion of a conical pendulum and states that the bob moves in a circular orbit of ever decreasing radius and that the string inscribes a conical form of ever decreasing volume. It is also exemplified by a student who watches a time-lapse movie of the life cycle of an organism and who then summarizes what he has seen as a sequence of growth stages. *Grouping* objects using immediately *given common properties* simply means to group objects based on their characteristics. To perceive that a mixture of algae observed on a slide through a microscope contains some filamentous kinds and some spherical kinds is such an instance of grouping. *Grouping* of objects based on *inferred function* or *inferred properties* is a higher-order observation process that borders on the one I call *cognition.* Through observation of a set of objects one may infer their respective functions and then group together those with similar inferred function. On observing a radio electrical curcuit for the first time, a student may infer that the electron tubes have a different *function* than the capacitors; he thus separates a total set of objects into two groups. An example of grouping based on inferred *property* is the separation of rocks into groups based on their inferred differences in physical density or chemical composition. (Such groupings may involve or lead to *ordering.*) These inference steps may require an interpretation step in scientific inquiry before the perceived groups are formed. A discussion of higher-order inference of form and function will be presented in the section on *interpretation.*

Cognition.   The next step in observation beyond *attention* and *perception* is *cognition.* Cognition is a process of mentally transforming or otherwise mentally manipulating that which is under observation. It is conceived here as a very primitive form of assigning meaning to experience by representing experience in various perspectives or conditions. As such, it is clearly different from relating current experience to prior experience through memory. That process I call *integration.* Integration is the next higher-level process category in Dimension 3, and is a logical extension of its first category – observation.

Cognition, the highest-level process of observation, involves the ability to represent objects in various altered or transformed ways. The mechanism for such cognition of form may be a mental spatial visualization or a mental symbolic representation that allows reorganization of the percept.

Such reorganizations are of two kinds. *Form perspective* involves visualizing an object in various positions vis-à-vis the observer. *Form transformation* is mental manipulation of properties of the object itself; for example:

visualizing changes in size or color of an object, or changes in shape (as produced by folding or compressing), or mentally enlarging microscopic elements of an object as though isolated from the whole, or visualizing opaque forms as though they were transparent, or seeing a sphere and imagining that it is hollow or that it is solid and composed of concentric layers or that it is filled with a spongy network.

Guilford[2] has constructed a classification of mental activities that pertains to our discussion of observation in general and cognition in particular. He has identified certain basic mental products that result from cognition. He defines cognition as an awareness, immediate discovery, or rediscovery of things in the environment. His use of the term *cognition* is consonant with the way it is used here; therefore some elaboration of his views may enhance the meaningfulness of the discussion of cognition and extend the concept to include other categories.

Guilford identifies six products of cognition: (1) units, (2) classes, (3) relations, (4) systems, (5) transformations, and (6)implications. Cognition of *units* means identification of relatively segregated or circumscribed items of information. This is consistent with the usage presented here. A unit is an individual object or form segregated from the surroundings. It is the equivalent of the Gestalt psychologists' "figure against background." According to Gestalt theory we tend to segregate objects from the surrounding field of sensation. When we look at a horizon, for example, we tend to segregate objects as individual entities from the surrounding space represented by the horizon and the sky in so far as it is homogeneous to our vision. Cognition of units is therefore a comprehension or awareness of individual forms. There is the implication here that one also abstracts some sense of identification for the item, as by placing a name upon it.

Cognition of *classes* is an awareness of groups of things that by presence of common properties or other similarities can be placed together. This category is akin to what we have called *grouping* in *pattern perception*. It requires that an individual be aware of the properties of units so that they can be grouped together based on their commonly possessed characteristics.

Cognition of *relations* is an awareness of a connection between two things based on variables or points of contact that apply to them. Relations include quantitative connections such as those represented by "greater than," "less than," and "equivalent to" statements. Qualitative characteristics such as relative shades of color, shape, or surface quality are also included here as connections among observed things.

A *system* is an organized or structured aggregate of items that is composed of interrelated or interacting parts. The concept of system is used widely in the sciences and therefore is a worthy addition to our list of things

2. J. P. Guilford, *The Nature of Human Intelligence* (New York: McGraw-Hill, 1967).

that can be observed or comprehended. The difference between a class of things and a system is that a system is a group of interacting components that have some mutual influence upon one another whereas a class of things is simply a group of things combined with one another based on common characteristics. In a system, the interacting components may have structural influences whereby each part gives strength or organization to the other component and in turn is strengthened by it. Or there can be functional influences whereby components interact to effect a certain outcome. That is, a dynamic sequence of events is maintained or initiated through the interaction of components. For example, the internal skeleton of an animal is a system of interconnected parts that by their spatial arrangement and articulation provide a supportive structural framework for each other and for the other parts of the animal. The circularory system in addition to its structural features also produces blood flow and distribution to all parts of the animal body and hence has a functional significance. In physics, an atom can be considered a system, since the nucleus and electrons are organized in such a way as to yield a stable yet dynamic organization of interacting protons and electrons. A system must be more than a collection of things. It must contain a set of interacting components.

A *transformation* is a mental product that is a rearrangement or reorganization of information. In its simplest form, it represents visual transformations of the kind we have described above, but in more complex instances it is a mental operation that produces a different expression for an observation, as for example by describing an observed action in mathematical or symbolic forms. This higher-level expression comes closer to the category called *interpretation* in this book.

Implications in Guilford's scheme are logical deductions or anticipations that can be drawn from experience. They are close to what we have called *inferences,* as in the case of inferred form or inferred function as a basis for *pattern perception.* In its highest form, however, it can include prediction, which is a higher-order category in the model presented in this book.

In many respects, Guilford's use of *cognition* is broader than mine. Moreover, I have chosen to separate various degrees of cognition into certain subcategories of Dimension 3 rather than lumping them together. Guilford's classification provides yet another view of the kinds of mental processes that students can be encouraged to acquire during a laboratory experience.

Integration

Integration is a cognitive process of giving meaning to experience by making simple associations between present experience and prior experiences. We make a first-order association of this kind when we say (or think), "The object now being observed is like one previously observed." At a higher level of integration a person realizes that an object is an instance of a class of objects *previously learned.* In order to realize that a presently observed entity belongs to a previously learned class, one must know the defining characteristics of instances that belong to the class. If a student is exhibiting difficulty in relating current experiences to prior ones, particularly in recognizing the instances that belong to a group or class of objects, some help is needed in clarifying what constitutes class inclusion and what kinds of characteristics (properties or attributes) of an object allow it to be assigned to a particular class. More will be said about this in Chapter Four. When an individual is able to identify an object or event that is only partially like previous experienced instances, abstracting out the peculiar properties that make it an instance, this is a very high level of integration. For example, a student knowing the characteristics of mammals and fishes, and observing a dolphin for the first time, is exhibiting a high level of integration if he correctly states that the dolphin is a mammal, not a fish. Such a high level of integration approximates the next level in Dimension 3 – *interpretation.*

Interpretation

Interpretation is defined here as a process of generating explanations for observed phenomena by using integration to associate immediate experience with prior-gained knowledge and applying logical reasoning to reach conclusions. This is a much higher level of cognition than integration. It subsumes integration and in addition requires use of logical reasoning to generate explanations. Although there are many cognitive processes that can be discussed under this category, we will consider the following five, selected because they are particularly useful in analyzing student behavior during laboratory learning: (1) unit explanation, (2) pattern explanation, (3) classification, (4) correlation, (5) causal explanation.

Unit Explanation. In unit explanations, a perceived unit (single entity identified in the environment) is analyzed as to its function or role in a natural system. This skill of interpretation requires that an individual identify an object or a set of objects and render a judgment about their function in the system under analysis. As an example, a student who has opened an incision on the dorsal anterior surface of a live anesthetized earthworm can observe the pulsations of the aortic arches surrounding the esophagus. Some careful reflection on the kinds of activities normally found in living animals may suggest a pumping action, and the red color may further suggest blood; thus a logical deduction would be that this is a simple heart. There is no requirement in this category that the student exhibit an understanding of the organizational relation of the heart to the remaining part of the circulatory system. Some care in teaching is indicated to help students focal-attend to unit objects and make interpretations.

Pattern Explanation.   In this kind of interpretation, an organizational principle or plan is deduced from observation and the significance of the organization is explained. A set of units is identified and a pattern is recognized among them. The functional significance of the pattern is also explained. Consider again the case of a student observing a dissected earthworm. The heart has been identified as a unit; a set of dorsal vessels is identified; an esophagus and crop are identified. The student can detect a pattern; namely, that the circulatory system consisting of heart and blood vessels spreads as a net over and around the tube-like digestive tract composed in part of esophagus, crop, gizzard, and intestine. Some visual tracking is necessary to conceptualize the organization of the circulatory and digestive systems. The pattern of relationships between the systems is deduced by grouping based on spatial relation. The student may deduce that the function of the netlike circulatory system is to gather and distribute materials within the body of the earthworm.

Two mental constructs often included in pattern explanation are inferred form and inferred function. *Inferred form* is internal structure deduced by examination of external shape or other cues, perhaps topographical or auditory. As an example, a student observing the circulatory system of an earthworm can infer, from the action of pulsation and general response to touch, that the blood vessels and heart arches are hollow. Similarly, one can infer that the earthworm body is partially hollow inside on the basis of the tactile data of gentle compression and by observing the movement and wrinkling of the body upon contraction. A geologist can make an inference about subterranean stratigraphy by examining the hills and valleys of the surface terrain. Processes of verifying such inferences and other kinds of interpretations will be discussed in the section on *evaluation. Inferred function* is a role or an activity deduced from some superficial data. For example, a physician makes an assessment of heart activity based on heart sounds (through application of a stethoscope) or an EKG record. A geologist observes flow of lava and flame from a volcano and makes some assessment of the action taking place underground.

Classification.   This mental activity, a process of *interpretation,* is more than merely placing an instance of a class into the proper classificatory niche through the use of memory. I have called that a process of *integration.* As defined here, "classification" means development of or addition to classification schemes. When one develops a new concept based on observation, this is an instance of interpretation through classification. To develop a concept, one must assign a name to the new conceptual class of instances and state the defining characteristics of the instances to be included in the class. Such attributes as physical properties of color, form, weight, and so forth can be used to denote the kinds of objects to be included in the conceptual class. Another kind of classification is the development and application of classification systems such as taxonomic keys and other categorical systems for ordering data.

Correlation.   An interpretation of phenomena that establishes an ordered relationship between two or more variables is a correlation. There is no assumption that one variable is the cause of another. Correlation is an explanation of relationship that states how two or more variables vary with respect to one another. A positive correlation is an explanation that one variable increases as another variable or set of variables increases. This does not allow us to assume that these variables are linked in a cause-effect relation, since it may be that both of the observed variables are responding in the same way to a common causal factor otherwise quite separate from them. Increased heart beat and increased breathing rate may be observed to be positively correlated. Neither is *necessarily* the cause of the other. Heart and lungs may both be responding to an elevation in body temperature. It is very important to help students realize that correlation is not sufficient evidence to suggest causal relation. Negative correlation is a relationship between variables where one increases as the other decreases. This also does not permit us to state that there is a causal relation.

An approximate estimate of degree of correlation between two variables can be obtained by graphing the response of one variable on the ordinate and the response of another on the abscissa. The plotted points in a scatter diagram indicate the direction of correlation. If the general slope of the line representing the points is positive (going from lower left to upper right), the correlation is positive; if the slope is negative (from upper left to lower right), the correlation is negative.

Causal Explanation.   This process of *interpretation* is the identification of certain ordered relations among two or more variables where sufficient evidence has been obtained to state that one variable or set of variables is a source that produces change in another variable. Cause-effect explanations have been widely used in science, and very frequently the term "cause" is used in the vernacular without much thought about its significance. Considerable imprecision in thought and language results from improper use of the concept of causality. Science learning is an excellent opportunity to develop greater sophistication in the use of causal thinking. Some basic principles to be used in developing causal explanations are presented here.

To establish that one variable (X) is the cause of another (Y), it is necessary to gain evidence that the supposedly causal variable (X) is the *only* one that plausibly could cause the event (Y). There are certain requirements that must be met. If X is a cause of Y, then X must precede Y in time; obviously X cannot be a cause of Y if X occurs after Y occurs. If X and Y are causally linked, then they should vary concomitantly. That is, a variation in X should produce a consistent variation in Y. If, for example, X increases, then Y should increase; or if X decreases, then Y should decrease; or some other regularly occurring pattern of coaction should be evident. Finally, if X is to be a cause of Y, we must be able to say with a degree of confidence that all other possible variables that could be causing Y have been controlled or otherwise accounted for so that we are left with a strong conclusion that X and only X is the source producing changes in Y. We normally use controlled experiments with matching groups that differ only with respect to the suspected causal variable X as a means of controlling for other possible causal linkages. It is always legitimate to ask to what extent an experiment has been sufficiently well designed to rule out other variables than X as a causal source. One must also remember that the context of a test of causal relation can influence the degree of effect produced by a cause. For example, in a chemical reaction where X is suspected of being a cause of Y, the temperature of the reaction vessel can influence the magnitude of the response observed. There are contributory variables that are not considered to be part of causal relation but tend to enhance or depress the magnitude of the relation.

Some consideration needs to be given to the concept of "necessary and sufficient conditions" in finding causal relations. A *necessary* variable is any variable whose presence is required if an event is to be observed. This does not mean that there is an automatic or definite assurance that an event will occur when the necessary variable is present. Rather, a necessary variable is one that is required in order to have an event occur. A *sufficient* variable is one that definitely leads to the occurrence of an event when it is present. But scientists are becoming increasingly aware that the idea of a single cause for an effect is probably not realistic and that it limits our comprehension of the complexity of patterns of cause and effect that occur in nature. Consequently some attention has recently been given to the concept of *multiple causation*. Two or more variables taken collectively may produce an effect in one or more other variables. As an example, death of a fish population in a pond can be attributed to eutrophication. But when we investigate eutrophication itself — excessive growth of organisms in the water, leading to overpopulation and depletion of the oxygen supply — we find it is usually a result of the combined action of several nutrients in overabundance under certain conditions of light and temperature. It is wise to help students progress from thinking about unitary causal relations toward thinking about patterns of causes yielding certain effects. In biology, interpretations of ecological, physiological, and animal behavioral events for example are likely to involve multiple-cause phenomena. And in the physical sciences also, including astronomy and electrodynamics, we can anticipate finding multiple-cause explanations.

Prediction

The term "prediction" is used in common language to mean a prophetic statement of something that is going to occur in the future. It is used in this book with a much broader meaning. To make a *prediction* is to make a statement about an anticipated occurrence. The occurrence can be a future time occurrence, in which case the prediction is indeed a temporal one (prophecy), *or* the occurrence can be a relationship between variables that one expects may be found to exist. A statement of hypothetical cause and effect is considered to be a prediction, as is also the statement that a phenomenon is expected to occur at a later time. There are three kinds of prediction that we will consider here: (1) pattern prediction, (2) extrapolation, and (3) hypothesizing.

Pattern prediction is a process of extending a known set of relations to include new ones. For example, given a food web, a student may "predict" that another organism should be included within the network. The prediction that a hitherto unidentified planet exists — deduced from certain observed perturbations in the orbits of known planets — is also an instance of pattern prediction (pattern extension).

Extrapolation is a special case of pattern prediction. It is the extension of a set of data beyond that already obtained. Particular emphasis is given here to extension of trends and sequences, as in predictions made by extending a graph or by projecting a sequence of events beyond the present trend. A student may observe an ecological succession in progress and predict its future course.

Hypothesizing is stating a relationship between independent and dependent variables, with due attention to contributory variables that are to be controlled during the test of the hypothesis. An independent variable is one that is to be manipulated by the experimenter or otherwise carefully observed and recorded; a dependent variable is one that is to be observed in relation to results of variations in the independent variable. Here is an example of a hypothesis: Adrenalin administered to the heart of a resting animal will cause increased cardiac frequency. (The independent variable is the amount of adrenalin administered, the dependent variable is cardiac frequency, and the control variable as stated is that the animal must be at rest during the treatment.) Testing of hypotheses is discussed in the next section.

Evaluation

Evaluation is a mental process of making a judgment about an experience with reference to a criterion or standard. The criterion used should be appropriate to the class of experiences one is judging. There are various kinds of criteria; for example, standards of excellence, moral standards, and observations that stand to confirm or negate inferences or predictions. As a result of applying a criterion, a judgment is rendered about the adequacy or quality of the experience to which the standard was applied.

In discussing pattern explanations as instances of interpretation, I cited *inferred form* and *inferred function*. One of the simplest kinds of evaluation can be verification of such an inference. It is not always methodologically simple, but relative to some other kinds of evaluation to be discussed later, it is *psychologically* simpler. Inferred form can sometimes be tested by making appropriate dissections or excavations to evaluate by direct examination whether the inferred form exists. The criterion in this case is whether or not the spatial relations expected to exist are indeed present. In some cases it is not possible to make direct observation of the phenomenon inferred, and one must use criteria of multiple sources of inferential evidence to yield a conclusion as to how confident one can be about an inferred state. For example, we may interpret a hill to be covering a rock dome underneath. If we cannot directly drill into the hill or excavate part of it, then we may choose to make more inferential tests, as by investigating how rock layers beneath the hill reflect sound waves, and by examining the surrounding terrain for other kinds of evidence. The same kind of reasoning applies to evaluating inferred function. If we cannot directly observe or manipulate functional components to evaluate an inference, then we must use sources of indirect evidence. When we use multiple sources of indirect evidence, we are usually forced to concede that our conclusions must remain probabilistic: more so even than when direct observation is used. Therefore it is useful to state how confident one is about inferred form and function. This statement will depend on how many indirect indicators have all yielded the same conclusion.

Evaluation of *predictions* varies in complexity depending on the kind of subcategory we are dealing with. I will discuss evaluation of pattern predictions, extrapolations, and hypotheses.

Evaluation of pattern predictions and extrapolations can be performed in a rather straightforward way. We observe whether the anticipated extensions are realized. For example, we observe natural phenomena to determine whether a predicted sequence of events or enlarged set of relationships actually does occur. We admittedly have to use various degrees of caution to be certain that we are adequately and precisely examining the situation to be evaluated. The criterion is confirmatory immediate sensory data. If the sensory data confirm the prediction, we accept it as valid. If not, we may try to determine, by careful analysis of the data, whether it has been fully and/or irrevocably rejected.

Evaluation of hypotheses requires the use of appropriately designed tests to determine whether the stated relationship among variables is actually observed to occur under adequately controlled conditions. Under adequate experimental conditions, one must be certain that only the independent variable is being varied and that other variables are controlled or adequately accounted for. Evaluation can include application of appropriate statistical techniques to determine the degree of confidence that we have in the findings.

Evaluation also permits making judgments about the quality or desirability of scientific findings from an *ethical* viewpoint. Here, the criteria applied are standards of what one considers to be valuable, desirable, or otherwise beneficial. We must recognize that some value systems clearly originate from fields other than science. To that extent, evaluation using these systems is outside the realm of science. One should keep well in mind what value systems originate within the scientific community and which ones originate from other sources such as general society, religion, or philosophy. The latter systems may be useful in evaluating aspects of science and therefore can be applied in making value judgments, but we must clearly know when we have exceeded the normal limits of the norms and values that pertain directly to scientific inquiry. These norms inherent in scientific communities have traditionally centered on the values of honesty, precision, creativity, and significance. We will explore this category more fully when Dimension 3 is considered in combination with Dimension 1.

Reconstruction

Evaluation of an experience with respect to scientific standards or newly acquired information may necessitate changing old models to better accommodate the new data. When a hypothesis derived logically from a theory is not confirmed under conditions of an adequate test, then one may need to reexamine the theory to determine if it is sufficiently valid in light of the new data. If evidence is obtained to show that a change in conceptualization is desirable, then the theory may be modified to accommodate the findings. This process of expanding, revising, or otherwise "correcting" a theory in response to new data is called *reconstruction*. The amount of change required in the theory should not be so great as to suggest that the theory be discarded or completely revised. Rather, the data should suggest alterations in the theory. This process of reconstruction is one of the more advanced kinds of cognitive functions in Dimension 3. It is not reasonable to try to deal with the complex psychological factors that may be required to reconstruct a theory. There is so much creativity and divergent logical analysis required that it is beyond the scope of this book. Indeed, so few psychological data are available on this specific process that much of what could be said would be highly speculative. The purpose of this discussion is simply to call attention to the possibility of this kind of behavior; very able students can be encouraged to examine theories to determine their adequacy. A review of Margenau's model of reality as presented in Chapter Two should be helpful in understanding the epistemological bases for alterations in constructs.

Construction

The production of a new theory or model to explain phenomena is called *construction.* Obviously, few ideas begin *de novo.* Most theoretical statements in modern science have some antecedent. This is particularly true in fields of science that have existed for some time. Therefore, when one speaks about the production of a new theory one is making a relative statement. A new theory is a combination of abstractions in an obviously novel representation of phenomena. The new statement represents an abstraction sufficiently unique that most people would consider it to be more than a major revision of an existing theory. The new view may result from discarding an older theory that is no longer sufficiently comprehensive or that is not capable of yielding verifiable predictions. Few students in our average classes will be able to exhibit the sophistication of creativity and intellectual skill necessary to achieve this kind of behavior. Some students may be able to generate novel models of natural phenomena that at least mimic some of the characteristics of the more sophisticated kinds created by the professional scientist. The student-generated models will probably be less rigorous in logical design than those of the scientist, and undoubtedly less general, less elegant. Very few truly elegant theoretical statements emanate even from the professional science field in any limited period of time (say within a decade); it is clearly unreasonable to expect that students will be able to achieve such accomplishments.

Approximation to scientific theory building, however, is possible for some students. Simple explanatory models of why animals behave in a certain way, or why plant or animal communities observed on ecological field trips occur as they do, are at least moderately respectable approximations to the behavior of scientists. To the extent that these moderately abstract models can be used to explain new instances or to generate hyphotheses for testing, the student will have achieved a respectable appreciation for the work of theory construction and evaluation. If a student cannot create a model of experience, he at least should be introduced to some uniquely respected historical exemplars of scientific theory construction. The field of physics is particularly rich in novel explanations that have been of revolutionary proportion. And in biology the Darwinian theory of evolution was, at its inception, sufficiently novel to reorder the thinking of scientists and laymen alike. Attention to the historical contexts for the origins of such theories is needed, in addition to exhibitions of their capacity to comprehensively explain phenomena and yield predictions to be tested.

Category Combinations

Assembly of the three dimensions into a graphic model as in Figure 5 helps us think systematically about combinations of categories from two or more dimensions. The kind of student behavior indicated by these combinations is more enlightening than considering each category separately. Moreover, by keeping the model in mind as a guide to analyzing student behavior, you can detect patterns of information-processing that might not be visible otherwise. Some students may consistently exhibit behavior in the laboratory that is very compulsive or restricted to only a few category combinations. When this is recognized, assistance can be given them in expanding their repertoire of behaviors.

Relations between categories in Dimensions 2 and 3 will be explained first; later, combinations of categories from all three dimensions.

The pattern for combining categories in Dimensions 2 and 3 will be to identify a row category in Dimension 2 and systematically relate it to column categories in Dimension 3. As a shorthand expression of the combined categories, the row category will be cited followed by a slash and the column category. Thus, sensation/observation will mean a combination of sensation and observation. This is the first row-one category pair to be considered, and sensation/construction is the last. Selected category pairs will be used to illustrate the kinds of interpretations derived from the model (Figure 5).

The combination *sensation/observation* represents the kind of sensory modality a student uses in performing observations and the degree to which a modality is used with precision and sophistication. Analyses of student behavior indicated by this category are determining the range and specificity of modalities used in observation. Does a student tend to fixate on one sensory source of data without due attention to other forms of data? The ability to identify both patterns and units through different kinds of modalities and careful discrimination among various kinds of unit percepts indicates a sophisticated use of sensory reception. (Some factors involved in sophisticated use of sensory modalities are presented later, in the discussion of operation/observation.) Sensation/observation is the most rudimentary combination of categories among all those within the model. It is a basic combination representing the kinds of responses we make to immediate sensory data.

The combination *sensation/integration* represents the kind of sensory data one calls up from memory to associate with immediate sensory experiences. As with the previous combination, we are concerned with the range of sensory data one has stored and can retrieve and the facility with which one makes associations between immediate experience and prior experience.

In a similar manner, the combinations sensation/interpretation, sensation/prediction, sensation/evaluation, sensation/reconstruction, and sensation/construction represent the use of sensory data in the remaining science process categories. For example, *sensation/prediction* would mean using various kinds of sense-derived data in making extrapolations, extensions, and hypotheses. Questions to be asked: To what extent does the individual fixate on one or more sensory-based variables, or to what extent is he flexible and diversified in his selection? What are the boundaries the student places on the extrapolations he makes with a given kind of sensory data? Is he overly conservative, failing to see the full range over which it is reasonable to extend a trend, or does he make extravagant extensions beyond what is warranted by the prior-gained data? In connection with the combination *sensation/interpretation*, the kinds of sensory data used in creating explanations, as well as the care and precision used in organizing various kinds of sensory data, are considered. Some students may be particularly conscious of and adept at explaining visual data but not have equal facility in generating explanations of aural, olfactory, or tactile data. Again, as with prior categories, the teacher should be ready to encourage more diversified use of sensory data in making interpretations of experience. The general principle used in interpreting combinations of sensation with science processes is the kind and range of sensory data used with each process in Dimension 3.

We consider next the binary combinations of *predisposition* with particular science process categories. The *predisposition/observation* combination represents a tacit awareness of and confidence in one's ability to use observation processes. For example, a student may exhibit an inclination to identify patterns but may become very timorous when confronted with a problem requiring careful identification of particular instances. Predisposition states are by definition non-verbalized; one must deduce these mental sets by observing the end product of a student's behavior. The other binary combinations of predisposition and the various science processes also indicate an individual's tacit inclination to be aware of, and confidence in his ability to perform, the processes. There is no requirement that the individual exhibit performance capability with the processes.

The binary combinations of the *recognition* category with science process categories have to do with the extent to which an individual can verbalize an explanation of the purposes of, and steps to be performed in, a scientific process. For example, the combination *recognition/interpretation* involves the extent to which an individual can describe how a classification system is constructed and used to organize and interrelate sensory data. *Recognition/construction* represents an individual's ability to describe the organization of theory and how theories and models of natural phenomena are constructed. There is no requirement that the individual actually construct a theory or model.

Each combination of the *operation* category with a science process category represents the extent to which an individual exhibits performance capability with the given science process. For example, the combination *operation/observation* includes an individual's ability to gather sensory data by sustained use of sensory modality; here is also included the kind of organization identified in stimuli by the observer. The tendency to consistently identify complex patterns or to identify only unitary or fragmentary instances of phenomena is contained here. The difference between unit and pattern identification in observation is illustrated by the following example. When presented with a survey map of a river basin area adjacent to a mountain having tree growth over the entire area up to a timberline on the mountain, a student may identify only instances of kinds of trees and herbaceous plants growing in the range. This is an example of unit identification. An example of pattern identification would be a statement that softwood trees grow along the river edge and grade into hardwood deciduous trees near the mountain base followed by a gradient into conifer trees up the mountain side toward timberline.

While discussing the sensation/observation combination, I introduced the concept of *sophistication* in use of *observation.* This will be further elaborated here. One of the outstanding characteristics of a maturing individual is his capacity to make finer and finer discriminations. This process, when particularly sophisticated, involves not only being able to observe differences in things but also being able to know when one kind of behavior is appropriate and when it is not. It also includes the ability to combine behaviors in appropriate ways to meet requirements of a problem situation. Behavior occurs within a context of many stimuli. Most social and physical settings in which we operate are composed of many components. In order to respond effectively in such a situation, we need to be able to abstract the various significant components and adjust our behavior to be consonant with them. Observation in the laboratory is a phenomenon of this kind. There are appropriate times to use a certain modality for data gathering and other times when it would be inappropriate. One role that an efficient teacher can perform is to help students better determine when an observation skill is appropriate. Careful teacher observation of a student who chronically fails to gather rational conclusions may reveal that he has not developed a fine enough discrimination as to when unit perception or pattern perception is appropriate. He may also not have sufficient acuity in a certain sensory mode. *Operation* indicates both a verbalized and performative knowledge of a science process. Therefore it is legitimate within discussion of this topic to consider the student's verbalized awareness of the ways an operation can be modified to better achieve an end result. Dialogue between teacher and student, or among students, about the rationale in using an observation skill can help to enhance sophistication in its use.

Similarly, the category pairs combining *operation* with integration, interpretation, prediction, evaluation, reconstruction, or construction represent a student's ability to perform the skills and to know under what circumstances they are appropriate. These combinations represent the highest-level mental activities in the model, the pair operation/construction clearly being of great abstraction and complexity. As a final clarification of binary categories, illustrative examples of two other operation combinations will be contrasted.

To illustrate *operation/interpretation* where the student employs a pattern explanation, we will use the same geographical range problem as set forth in discussion of operation/observation. The student produces a statement that softwood trees grow near the river's edge since they are mesophytes well adapted to water-laden soil and require abundant water to live, whereas conifers grow on the mountain slopes since they are wind- and drought-resistant due to leaf mechanisms that prevent excessive water loss. This instance of operation/interpretation is to be contrasted with *operation/integration.* Instances of the latter indicate that the student is able to identify prior-gained instances of the observed pattern or units and to state the conditions of relationship that exist between prior knowledge and the newly acquired data. For example, to state that a vegetative growth pattern presently observed in Nebraska was previously seen in a grassland in Idaho is an instance of integration. The simple classification of upland conifer trees as belonging to the class Gymnospermae is also an operation of integration.

We proceed now to a survey of ternary combinations among categories in Dimensions 1, 2, and 3. Combining Dimension 1 categories with those of the other dimensions allows us to include human value systems, norms, and beliefs as components of the scientific enterprise. The use of Dimension 1 can help us make value judgments about the impact of science on society and on the natural environment, and about the role of the individual in performing scientific inquiry. Selected exemplars of ternary category combinations are presented. We use the terms *nature* and *society* to stand for the two Dimension 1 categories: *natural environment orientation* and *social environment orientation.*

Consider the combination *nature/predisposition/observation.* If an individual views the natural environment as being chaotic or controlled by capricious forces beyond the comprehension of man, then there is a low probability that the individual will develop a predisposition to identify order in nature. There is little likelihood that he will have confidence in his ability to gain ordered information from sustained observation.

A full understanding of the predisposition category is useful in preparing science curriculum experiences, particularly for those students who have lived in social settings where experience is believed to be capricious or where it is believed that natural phenomena are governed by chaotic or wholly incomprehensible forces. Youth from low-income backgrounds or from social milieus where superstition prevails often lack confidence in their ability to order sensory data and to construct systematic explanations of natural phenomena.

There are some interesting ternary combinations involving the social environment orientation category. Assume that one holds, as a value judgment, that the environment is to be used for social technological advancement without regard to environmental impact. Then consider the combination *society/operation/evaluation*. Suppose the results of a scientific experiment show that addition of nitrate to a depleted cultivated field will enhance crop production but that run-off will undoubtedly lead to eutrophication of a nearby estuary which is a central breeding ground for a species of fish highly critical in the food chain. The results of the experiment can of course be evaluated first in terms of the accuracy of scientific design. But in addition, the use of Dimension 1 suggests an evaluation in terms of the impact of the application of the findings on the ecosystem. Criteria can be discussed as to what constitutes sufficient technological gain and relief of human needs to permit a certain level of environmental insult.

The general principle to be remembered in combining Dimension 1 categories with those of Dimensions 2 and 3 is that the binary combinations previously described are now further modified to take into account the students' belief systems about the social and natural environments. Classroom discussions preceding and/or following a laboratory experience will frequently reveal these. If a student raises an environmental question sufficiently appropriate to the topic at hand, the class should be encouraged to follow through on a discussion of its relevance to the scientific topic. If it is inappropriate or irrelevant in that context, then this should be stated with as much candor and consideration as possible.

Science Process Sequences

The practice of science is notable for its creativity in inventing and applying methods for obtaining information about the natural environment. For this reason, it is impossible to say that the sequence in which science activities were presented in our discussion of Dimension 3 is invariant or even optimal. That sequence represents a psychological gradient from those experiences which are most immediate toward those which are more abstract and symbolically representational. It is probably true to say that during maturation human learning can progress optimally from the more immediate kinds of experiences toward those that are more abstract. Therefore it is probably safe to assume, as a general prescription for planning science laboratory curricula, that less mature students will benefit through beginning with experiences oriented toward observation and integration and progressively working toward the higher-level categories of evaluation and theory construction.

For a more mature student who has mastered several of the science processes, many possible sequences of their use are possible. That is, in any given laboratory experience, one may use several different combinations of science process activities in reaching a final solution. As an example, consider a student who has mastered the abilities of observation, integration, and interpretation. A laboratory task could require that he begin with observation to gain some basic data, then integrate this with prior knowledge, make an interpretation, and from this realize that more data of a particular kind are needed. Therefore, observation is used again to obtain the additional data which may then be integrated with the interpretative scheme previously developed. As shown in this example, recursion is one kind of pattern that can be used. By *recursion,* I mean going back to a previously used science process within the context of solving a given problem.

To further elucidate the kinds of process sequences that can be applied, I cite several illustrative examples. These are not necessarily the only patterns one can conceive. Rather they are exhibitions of particular kinds of sequences that seem likely to be suitable in commonly encountered laboratory experiences. Each sequence is cited and briefly explained. Some of the more fundamental sequences are presented first, followed by those that are conceptually more complex. In general, the more fundamental process sequences begin with the immediate sensory gathering processes and proceed toward more abstract processes. The more complex sequences begin with abstract processes such as construction and move toward less abstract ones. It should be noted and kept in mind that all of the processes are here represented at the *operational* level of Dimension 2.

A very fundamental sequence of gaining new information and making interpretations of it is *observation-integration-interpretation.* It is used when a student has an objective of making interpretations based on novel information not previously used. It also assumes that the student has little theoretical information to guide his initial observations. Such a sequence is used when a student is given an opportunity to synthesize interpretations fairly freely without contamination of predetermined conclusions. This is a kind of ideal situation that is rarely possible in practice. Most people bring some kind of theoretical or conceptual frame of reference to bear on most tasks. However, some are less clearly based in theory and can be properly described as beginning with pure observation and moving toward interpretations that are fairly free of contaminating prior expectations. An example of this kind of sequence is an open-ended laboratory experience where the student is allowed to create his own interpretation scheme. Consider the case of a biology student who develops a taxonomic key. He begins with direct observation of the specimens to be described, notes their characteristics, combines these data with prior experience (integration) to create classes of specimens, and develops a classification key based on his own observations (interpretation). As another example, taken this time from chemistry, consider a student who is presented with a novel chemical reaction. A solid when mixed with a liquid produces a gas. The student's task is to develop a means of interpreting the experience. This suggests that the initial observation leads to integration with prior experience to devise a way of collecting the gas and a test to determine the kind of gas produced and perhaps its volume. This sequence requires recursion; the initial observation is followed ultimately by a second observation to determine such things as gas volume and composition.

Another sequence is *interpretation-prediction-observation-evaluation*. In this case, an initial observation is not used. One makes an interpretation of previously gained data (perhaps from the literature), produces a prediction, makes an observation to gain data relevant to the prediction, and then evaluates the quality or validity of the prediction. As an example, a student examines some meteorological charts and thinks he detects a pattern of events that represents a relation between rainfall amount and the direction of travel of low pressure areas across a geographical range. He predicts that such patterns will recur with a certain relationship between low pressure, air movement, and rainfall. The student then examines subsequent meteorological charts (observation) to determine the validity of his prediction (evaluation). Evaluation frequently involves observation as a method of applying criteria to render a judgment. In this case of direct evaluation, the criterion is the actual presence of the predicted phenomenon as determined by observation.

The sequence *integration-prediction-evaluation* is a form of investigation commonly called an intuitive-based study. In other words, one does not begin with external sources of data to generate a research problem and design; rather, one integrates two pieces of information already stored in memory and from this yields a prediction that is then tested. *Integration* is presented here as an internal process of combining previously acquired data to yield a new understanding. This is a different sense of the term from the one presented earlier where we discussed the integration of newly acquired information with prior-gained information. The new use is consistent with the former. The difference is the sources of the data to be combined. In one case some of the data are newly acquired through observation; this is integration using an external data source. The second usage is based on recall of previously acquired data which are then integrated through logical analysis; this is integration using internal data only.

Another sequence type is *evaluation-interpretation-observation*. In this case, one makes a value judgment about a currently accepted procedure or interpretation in science and, if it is found to be lacking in some respect, one reinterprets the phenomenon and suggests new observations to be made. If the new interpretation involves changing a theory or general model in science, then of course the sequence is better presented as *evaluation-reconstruction-observation*. Whether it is reinterpretation or reconstruction depends upon the degree of abstraction and generalization of the explanation that is revised. An equally probable sequence of this kind (in this case recursive) is *evaluation-observation-reconstruction-observation*. Evaluation may suggest shortcomings in a prior interpretation which can be determined only by further observation. This in turn yields new data that suggest reconstruction of the model or theory. The new construction then suggests additional observations.

A sequence type that involves use of two abstract thought processes is *observation-reconstruction-prediction*. In this case observation produces some disparate data relative to previously developed explanations, suggesting a reorganized and perhaps a more comprehensive explanation than was previously formulated. The newly revised theory may include additional interpretations that can be tested (prediction). If the predictions are actually tested, then of course the sequence is extended as *observation-reconstruction-prediction-evaluation*.

A considerably more abstract sequence is initiated by constructing a new theory from prior-gained data not grounded in immediate sensory data. An example of such a sequence is *construction-prediction-evaluation*. In this case one synthesizes a new explanation, which yields a prediction, which is then evaluated through an appropriately designed experiment or observation. The initial step could equally likely have been reconstruction. In this case — *reconstruction-prediction-evaluation* — the guiding model would be a revision of an existing theory rather than the development of a novel explanation.

During the teaching of laboratory science, various sequential approaches can be anticipated and encouraged depending on the kind of laboratory problem presented and the ability of the students. Recurson is likely to occur frequently in some laboratory experiences, particularly with the use of lower-level categories in Dimension 3.

I trust that the foregoing discussion of our theoretical model is sufficient to lay a foundation for the more practical discussions, in the next two chapters, of how it can be used in laboratory teaching.

Four

TEACHING WITH
A NEW PERSPECTIVE

In the foregoing chapters we have discussed some conceptual bases for understanding laboratory teaching. In this chapter, we examine some of the practical implications of those discussions. We begin with some purposes of laboratory instruction. There are many reasons given in the educational literature for laboratory instruction. Not all contemporary and valid reasons can be cited here. We will consider four purposes that are most relevant to the perspective established in this book.

1. *The laboratory is a place where a person or group of persons engage in a human enterprise of examining and explaining natural phenomena.* The purpose of such an experience should be more than memorization of scientific facts and certainly more than confirming data presented by other instructional means. A purpose of laboratory instruction is to enhance a student's intellectual and aesthetic understanding of natural phenomena. The student should gain confidence in his ability to reliably gather and interpret sensory data, and to apply rational ways of thought toward explanation of natural events. This purpose places the psychological development of the student at the central focus of laboratory instruction. Whether planning to become a scientist or not, a student certainly can benefit from ability to deal rationally with natural events and from enhanced aesthetic appreciation.

2. *The laboratory provides an opportunity to learn generalized ` systematic ways of thinking that should transfer to other life problem situations.* The confidence gained and the rational and orderly processes of thought

acquired in laboratory work should be useful in broaching problems of daily life. This does not imply a "kitchen-oriented" science; what is desired is growth in basic processes of rational thought that can aid in various problem areas, both in pure science and outside it. Moreover, there should be transfer of the aesthetic responses developed in science instruction to one's appreciation of the environment, contributing to a positive orientation that will persist throughout life and give satisfaction to daily experiences.

3. *The laboratory experience should allow each student to appreciate and in part to emulate the role of the scientist in inquiry.* The degree of sophistication of these experiences should be commensurate with the student's abilities and mental maturity in the field of science.

4. *The result of laboratory instruction should be a more comprehensive view of science including not only the orderliness of its interpretations of nature, but also the tentative quality of its theories and models.* The dynamic qualities of scientific research, including the dramatic and sometimes revolutionary nature of new scientific conceptualizations, should be appreciated.

These purposes will be rendered in more operational terms through discussion of their meanings within the framework of our three-dimensional model.

Purposes and Functions of Laboratory Teaching

Purpose 1: To foster knowledge of the human enterprise of science so as to enhance student intellectual and aesthetic understanding.

It is clear from psychological literature that diversity in human abilities requires some consideration of the kind of content to be presented to students of various capabilities. The science process dimension of our model is a useful framework for considering ways of matching laboratory experiences to the abilities of students. We will build our main discussion around the categories in Dimension 3, considering the most rudimentary skills first and proceeding to the more advanced ones.

First, however, it should be noted that the aesthetic categories of Dimension 1 are closely related to the rational processes of science investigation and need to be presented and understood as part of the total psychological significance of such investigation. The student should be encouraged to think both about the emotional satisfaction aroused by perceived instances and about the scientific significance of the pattern in the natural phenomenon being observed. For example, a transmission radio station has a signal generating function that is *complementary* to the signal reception function of the receiving set. The impedance and inductance of the two circuits can be seen as complementary phenomena. Complementarity in this case enhances coupled responsing between the two systems. This is a particularly abstract example, from physics, of the remarkable generality of the aesthetic categories and the creativity one can use in applying them to various phenomena. I discuss connections between aesthetic properties and scientific significance more fully in the final section of this chapter.

We proceed now to relate the science processes of Dimension 3 directly to the purpose of fostering students' intellectual understanding of natural phenomena.

In any given class, regardless of grade level, the most basic science processes — observation, integration, and interpretation — will of course be used, and attention will need to be paid to building students' skill in their use. Some students, particularly those who come from culturally impoverished backgrounds or have grown up in a community where superstition was the source of explanations for phenomena, may need to have many very basic science experiences in which to practice the simpler skills before they can go on to more advanced processes. With culturally deprived youth especially, considerable attention may be needed to promote predisposition to these basic science skills and growing confidence in the possibility of making orderly interpretations of the environment. These students also need help sometimes in learning how to mobilize past experiences and associate them with present experience (integration). The teacher will need to be aware of their need for constant encouragement to scan memory for prior experiences and relate relevant ones to current experiences. A program of training in pattern identification will be necessary for some of these students. Their attention will need to be directed from the particulars that *are* perceived in natural phenomena to the general organizing schemes that *can* be perceived.

More advanced or culturally advantaged students who have already acquired considerable skill in the basic science processes may yet benefit from gaining verbalized awareness of the skills. Such a student can appreciate the teacher's assistance in putting word labels on the simpler science processes and in recognizing when unit identification and pattern identification are being used.

To able students the teacher should also point out the various forms of observation and interpretation skills and their characteristics in association with the various process orientation levels — sensation, predisposition, recognition, and operation. In the operation phase, an expectation of increased precision and sophistication in use of the skills can be emphasized. In recent years, while we have turned our attention in science curricula on global inquiry skills and open-ended experiences (perhaps even embracing the concept of "messing around in science"), we have sometimes lost sight of the clear contribution that science teaching can make to growth in precise and sophisticated ways of thinking. *Homo sapiens* is distinguished from lower forms of life by the intelligent use of precision in exploring the environment. Students who have been carefully introduced to laboratory and field science skills of observation, integration, and interpretation should have a better appreciation for their ability to think precisely and to know where certain kinds of interpretation skills are appropriate. Moreover, the aesthetic responses appropriate to science processes, rendered through identification in the environment of symmetry, balance, complementarity, and so forth, should provide a general mental set to regularly look for and recall these satisfying exemplars of man's orderly interpretations of nature.

In connection with prediction and evaluation and the higher categories, the teacher will need to exercise particular care in identifying those students who are already prepared to launch into activities requiring these skills and may be allowed to pursue them quite autonomously. Students who are less able will need specific help with the component skills involved. For example, in the case of predictions made by extending patterns or classification schemes to include new, not yet encountered instances, the student needs help in three phases: in identifying the general principles that give coherence to a category scheme, in identifying the relationship of individual categories in the scheme, and in looking for unidentified ones to extend the pattern. Suppose a student has learned a pattern of organizing plants into *hydrophytes* (growing in aqueous environments), and *mesophytes* (growing in moist soil). Some thought about this grouping will show that there is a principle of classification based on abundance of water in the environment. Hydrophytes are plants that grow in water or very wet soils; mesophytes are plants growing in soil that is neither very wet nor very dry. Logical extension of this classification suggests that there must be at least one more group: plants that grow on very dry land. It is this kind of thinking that should be encouraged in helping students extend patterns. (For students interested in language, this learning will be both facilitated and reinforced by knowing that *hydro-* means "water" or "wet" and *meso-* means "medium.") The final category in the extended sequence is of course the *xerophytes*.

The extension of graphs requires that the student fully understand what it means for two or more variables to be related by the function generating a graph, and when such a function is continuous and valid over the range of values to be used in extrapolation (if not, the graph cannot be extended).

Formulating hypotheses is particularly demanding. The student must comprehend the nature of causal relations as set forth in Chapter Three, be able to state a relationship between independent and dependent variables, and recognize the contextual variables that must be controlled in preparation for a test of the hypothesis.

Evaluation requires that the student be adequately informed about the use of criteria in making judgments. He needs to know that judgments are not to be rendered on impulse – that some criterion or standard must be applied to a situation in making a critical appraisal. Evaluating predictions requires additional care in determining the appropriateness of the criteria employed. For example, what criteria of research design must be applied? Should controlled experimentation be used, or some observational procedure?

The categories of reconstruction and construction require a high degree of logical and abstract thought. However, in building or reorganizing models and theories, a student also needs to be forcefully reminded that the concepts and logical relations he posits as explanations for natural phenomena must be rationally related to immediate sensory data. Each idea that he creates to represent an event must be logically linked to an observation that can be reliably assessed in its occurrence by two or more persons. It is useful to have students write their definitions of concepts and relations and cite the ulitmate sensory data that support these observations.

Consider the following example. A group of students are observing an aquarium containing a single species of fish both male and female. (The details cited in this example are not intended to represent observations of any particular species of fish.) The students observe that the animals behave in certain ways. They believe there are signals used by the fish to control group behavior: one set of signals related to schooling, another set to mating, and another to food-gathering. The construction of this conceptualization takes the form of a model of fish communication. The necessary requirement that the students have to recognize is that they must formalize the kinds of activities that the fish exhibit in terms of actually observable behaviors. For instance: What are schooling signals? They must first be defined conceptually, in terms of their functions ("Schooling signals are actions designed to orient the motions of the fish school, signaling forward motions, stops, or directions of motion.") Then the specific operational indicators must be described in relation to the functions. The result is an *operational definition.* For example, the students in the present case might arrive at this one: "Actions of the dorsal fin are the main signals for schooling; a raised dorsal fin appears to signal stopping a forward motion, a folded dorsal fin signals rapid forward motion, and curvature of the fin signals direction of motion." The mating and food signals of the fish can be similarly defined in operational terms.

From these precise definitions, various predictions based on the constructed model can be made about fish behavior. Since the behaviors are operationally defined, the predictions can be tested for validity. Communication between scientists is always more precise when operational definitions are used. Margenau's concept of constructed reality and the relationship between *P*-plane and *C*-field should be kept in mind and explained in appropriate terms to students who are engaging in construction or reconstruction of models and theories. Some advice can be given to students on ways of beginning to construct models. There are two ways, relative to the phenomena being explained: (1) inductive and (2) deductive. In the *inductive* approach, the specified set of phenomena to be explained are carefully observed and, with aid of general knowledge in the field, one identifies units or instances of events that appear to be significant. With these instances in mind one then synthesizes a pattern or set of relations among the instances which when fully integrated represents the inductively formed model. Consider again the fish communication example. If the students begin by observing the aquarium and build their ideas from rudimentary observations, this is inductive construction of models relative to the phenomenon. *Deductive* construction occurs when one uses existing knowledge in the field to generate new models without beginning with observation. In this case observation would follow the initial conceptualization to help clarify and confirm certain relations in the model. In the case of students building a model of fish behavior, a deductive model could have been constructed by consulting the literature on various forms of signals exhibited by aquatic organisms. Using this collection of data in the literature, a model of various patterns of behavior could be produced.

The variables so identified in a conceptual model can then be refined by new observations to confirm their appropriateness. Moreover, through actual observation of the events that correspond to the conceptual variables, the students can develop operational definitions for these variables. In a deductively derived model, observation serves two functions, it provides evidence for validation of the conceptual variables by use of immediate sensory data, and it facilitates precise operational definitions of the conceptual variables identified by the students.

The idea of *conceptual* and *operational definitions* of natural events is not an easy one to convey to students, for it presumes a general familiarity with the ideas of abstractness and concreteness which some students have not yet developed in a formal way. Therefore, some preparatory explanations by the teacher will be necessary to help the students comprehend the meaning of conceptual versus operational definition. One way is to use already established theories or models in science and present explanations of them, first citing conceptual definitions of variables and then immediately thereafter citing examples of operational definitions. A chart can be placed on the chalkboard with conceptual definitions listed on the left-hand side and their operational equivalents on the right-hand side. During this process it is useful to point out repeatedly that a conceptual definition is a generalized definition that does not use concrete instances. An operational definition is a statement of the concrete actions, events, or manipulations that are actually observed. Two pairs of conceptual and operational definitions are shown here as they might be entered in a chart.

Conceptual Definition	Operational Definition
A food web is a network of feeding patterns where predator consumes a prey and in turn is consumed by other predators.	In a food web animals and plants are linked to one another by the way in which they feed upon one another. An animal may eat a plant; the animal in turn is eaten by another animal; and so forth.
Energy is the capacity to produce work.	Energy is the ability of a stretched spring (formed from an elastic coil of wire) to move a cart to which it is attached when the cart is released from a resting position.

It is wise to have the students practice writing conceptual and operational definitions before they begin constructing a model. This can be done during the preparatory phase described above, allowing the students to extend the chart begun by the teacher.

Purpose 2: To foster science inquiry skills that can transfer to other spheres of problem solving.

If we were able merely to enhance a young persons's appreciation for science knowledge and methods of investigation, we would have accomplished a great deal. But most of us engaged in education believe that what is learned in our classroom should have significance beyond the teaching-learning event. In addition to developing richer understanding of the environment, we quite reasonably aim to provide some generalized ways of thought that will have significance in daily life. The ability to effectively use prior-learned skills in new settings is classified in psychological terminology as *positive transfer of training.*

We have come to know something about the conditions of learning that facilitate positive transfer of skills. To achieve postitive transfer, the conditions during the learning of the skill must have something in common with the situation in which the skill will later be applied. The more generalized the conditions in which the skill is learned, the more likely it is that the skill will have application in other situations. Bloom[1] has developed a model (see Figure 6) of the kind of mental functions that take place during transfer of a skill from one area to another.

[1.] B. S. Bloom (Ed.) and others, *Taxonomy of Educational Objectives, Handbook I: Cognitive Domain* (New York: McKay, 1956).

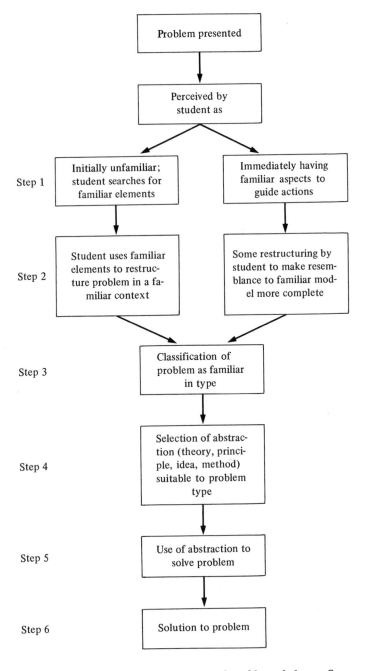

Figure 6.   **Mental functions in transfer of knowledge.**   Source: Benjamin S. Bloom (ed.), *Taxonomy of Educational Objectives, Handbook I: Cognitive Domain*, p. 121. Copyright © 1956 by Longmans Green & Co. Inc. Reprinted by permission of David McKay Co. Inc.

In Step 1 of Bloom's model, the student can classify the new situation as either (1) unfamiliar, thus requiring that he search for familiar elements to help him classify the situation as one that is familiar to him, or (2) immediately similar to previous ones he has learned, thus requiring little reorganization to make it suitable to guide action. In Step 2, the student reorganizes the elements of the problem to make them more consonant with his prior experience. If the problem situation was perceived as initially unfamiliar, then the student uses familiar elements to build an interpretation of the problem that will make it a part of a familiar context. If the problem is clearly like a kind previously encountered, then a moderate amount of restructuring is required. These steps are examples of the mental process of integration as represented in Dimension 2 of our model. The student is relating immediate sensory experience (the problem presented to him) to prior experience as a way of applying the prior learning toward solution of the problem. When the student achieves sufficient reorganization of the problem, he classifies it as a familiar type (Step 3). A principle or other kind of abstraction is selected to be applied to the problem (Step 4). In Steps 5 and 6, the principle is used to solve the problem.

The model has clear implications for teacher behavior in assisting students to apply prior learning to a new situation. A student may exhibit deficiencies in applying prior learning at any point from Step 1 to Step 5. A careful analysis by the teacher to determine what stage the student has reached can facilitate progress toward problem solution. A few suggestions by the teacher can help the student develop patterns of behavior that will facilitate future application. When a student consistently fails to negotiate a particular step in the transfer of skill, the teacher should help him recognize the kind of mental operations required to complete the step so that future encounters will be more successful.

A critical step in this processs of application is the identification within a problem situation of cues that will allow the person to adequately classify the problem as being of a familiar kind and hence amenable to solution by application of previously acquired appropriate responses.

There are two ways that psychologists have found to facilitate the critical step of finding familiar aspects in a problem situation: (1) the common-elements method and (2) the use of general principles to help the student identify general classes of problem situations to which a given skill can be applied. Each method will be explained.

In the *common-elements* method, we are able to specify in detail the kinds of situations to which a skill can transfer, and to identify the key characteristics of such situations. To this end, these specified key characteristics are incorporated into initial learning situations and are given sufficient saliency through pointing them out one by one, or by otherwise calling the students' attention to them, that the students will recognize them in a subsequent situation. The skill or other kind of learning to be transferred is taught in the presence of the situation elements and becomes associated with that context. When a similar situation arises (containing elements in common with the initial learning situation), the student can quickly recognize it as a type previously encountered and apply the learned response pattern to the new situation. As an example, consider the situation of training an individual to use the skill of applying a tourniquet to stop bleeding. The general set of elements in a situation that indicates use of a tourniquet is a cut that occurs on an appendage such as an arm or leg and bleeds so profusely that loss of blood may lead to hemorrhagic shock and death. A further refinement is introduced as to where to place the tourniquet on the appendage. If the cut bleeds in spurts, place the tourniquet between the body and the cut; if the cut bleeds in a steady flow, place the tourniquet at a potition on the far side of the cut, away from the body. The particular elements common to each kind of situation requiring a tourniquet are thus set forth, and under optimal conditions the student learns to apply the device at the appropriate place on the body while the teacher points out the specific elements.

This method of enhancing positive transfer has certain limitations. First of all, it will work only with those situations where the elements of significance are sufficiently obvious that they can be clearly enumerated and described. Secondly, the transfer tends to be very compulsive and to lack flexibility — the individual may not see an appropriate use of a skill in a situation requiring it simply because the new elements lack sufficient similarity to the originally learned elements.

To overcome these limitations of the common-elements method, the *general-principle* method is employed. In this method, the student is given a principle or set of principles that allows him to identify a situation that is appropriate for use of a skill. The principle is sufficiently general to apply to a wide variety of appropriate instances. For example, the compound light microscope can be used to observe objects that are sufficiently thin and translucent to transmit light. In chemistry, the ultraviolet spectrophotometer can be used to elucidate molecular structure of compounds containing double bonds. These principles state the general conditions where a manipulative skill is appropriate. There are of course intellectual skills used in conjunction with these manipulative skills. As an example of a principle that is used to mediate positive transfer of an intellectual skill, consider the case of inferred-form perception (subcategory under interpretation in Dimension 3). As a general principle in selective use of inferred-form perception, the student can be told that inferred form is to be looked for when it is impossible to make dissections, excavations, or other kinds of openings into an object to be observed. The principle gives an admittedly straightforward guide to the appropriate application of the skill.

One of the functions of an effective teacher is to help the student identify general principles and situational common elements that facilitate transfer of an acquired skill. As a student exhibits successful use of a skill, the teacher can help the student identify the various aspects of the situation that make the skill application appropriate. One of the limitations of the use of principles in mediating positive transfer of learning is that it can lead to overgeneralization in application. Sometimes the student is insufficiently aware of the limitations of the principle to know when a situation fails to meet the requirements of being an instance of the principle. In general it is helpful to use both general principles and the common-elements method in mediating transfer. The principles provide generalized knowledge to enhance student use of the skill in a wide variety of appropriate situations, while the common elements provide specificity to make clear what particular characteristics of a situation make application of the skill appropriate. In some situations, it may not be possible to identify specific elements and only a general principle can be cited.

Another point that needs to be made clear is that when we speak of transfer of training, we are talking about the *operation* category of Dimension 2. To speak about transfer of training from one task to another presumes that the individual can actually perform the skill to be transferred.

We are further talking about transfer of training from a science task to more general life situations. This condition suggests that Dimension 1 is also of importance. The orientation of an individual toward the natural and social environments in part determines how he will use cognitive skills in assessing evidence related to problems in these areas. To gain optimum transfer of training from the science laboratory to other life situations necessitates that the student be made aware of the generalized use of science process skills in other fields. The use of careful observation, in the various ways cited in this book, can be applied in numerous occupations, in making decisions relative to social and community planning, and indeed in engaging in problem-solving tasks in daily life wherever rational thought is required. The skills of interpretation and evaluation are clearly significant in overcoming tendencies toward prejudice and superstition. Obviously, one cannot devote excessively large amounts of time to these transfer functions in teaching. But it takes very little time during a discussion in the laboratory, or at a time following it, to point out that the same care one takes in demanding precise sensory data to make scientific interpretations should be used in making social and some moral decisions. Prejudice is still a grievous component of modern social thought. To the extent that science laboratory instruction can help young people learn to demand clear evidence before they categorize people, institutions, or beliefs, the more we can contribute as science educators to the general improvement of society.

While considering positive transfer functions in teaching, it is appropriate to discuss some particular methods of teaching the skills cited in Dimension 3. There are four approaches that can be used: (1) prescription, (2) prescribed selection, (3) induced selection, and (4) free selection.

Prescription of a skill means, as the name inplies, that the teacher states an appropriate bit of information or describes a skill that can be used at a particular point. It is clear that this approach offers the least autonomy for the student, but it has the advantage of helping him take a prompt and prudent step when the very success of the whole laboratory experience may depend on it. I do not advocate excessive use of prescription. Some educators may even decry its use altogether, as being too didactic. I take the view that a teacher must be very flexible in his approach to teaching. There are some students who need some support early in laboratory experiences, and teacher prescription gives them sufficient advantage to prevent a complete failure in the laboratory project. Hopefully, the teacher will gradually withdraw such massive support and apply some of the more indirect means cited below as the year progresses. Sometimes, when the student is sufficiently emotionally secure, he should be allowed to pursue the laboratory task with the possibility of error and failure. I do not believe, however, that a truly effective and conscientious teacher allows this to happen repeatedly. If it occurs too frequently, it means either that the laboratory tasks are simply too difficult for those students, or that the teacher has not given sufficient prior prescriptive training, or both. A student who consistently suffers failure or makes too many errors due to inappropriate tasks is very likely either to develop a negative self-image, hating science on that account, or to despise science itself, presuming it to be a failure-ridden or even imprecise field of study.

Prescribed selection of skills to be applied in a situation provides more student autonomy than does simple prescription. Here the teacher offers at least two skills and allows the student to make a selection. This provides experience for the student in making decisions based on the appropriateness of the method to the task. It also gives sufficient support to assist the student if he has not developed considerable prowess in making selections.

Induced selection is a method of indirectly getting the student to select a particular skill by asking questions that guide his thinking toward identification of appropriate skills. Selection can sometimes be achieved by asking the student if he can recall a prior situation like the current one and what he did then. This is an instance of inducing transfer of training from one science task to another. Or, the teacher may use a subtle sequence of prompting, by first asking the student to consider ways he might approach the task, and then, through dialogue, helping the student select and refine his interpretation of it and apply it. This approach and the foregoing ones require much teacher activity in the laboratory, much engagement with students in worthwhile conversation that yields cognitive growth. It certainly does not fit the all-too-frequently used method of the teacher who puts the students to a task and then declares "Hands off, let them just muddle through." The teacher can serve as a source of ideas that keep the students moving forward.

Finally, for those students who are sufficiently mature or who have been helped to mature through the successive approximations of moving from prescription to induced selection, there is the method of *free selection*. In this case, the laboratory task is set up in such a way that the tools provided and the kinds of objects to be studied facilitate student decision-making about skills to be used without teacher assistance. One creates a properly organized laboratory experience and provides sufficient opportunity for the student to gain background information such that the student can proceed with minimum teacher intervention. It requires considerable guidance and maturation before many high school students reach this stage of autonomy. It is better that we clearly recognize this and plan for *growth* toward autonomy rather than demanding too much too soon.

The uses of induced selection and free selection are consonant with guided and free inquiry teaching respectively. There is some evidence to suggest that use of induced selection and free selection in teaching science skills increases their meaningfulness, since the student must supply the mediating mental processes to select the skill and it thus becomes more clearly a part of the student's mental structure. Moreover, free selection can facilitate positive transfer to other situations to the extent that the student must work out the conditions of the situation in his own mind before arriving at the skill to be applied. If he has the proper preparatory mental set that the skills he acquires in science can benefit him later, he is likely to remember the conditions wherein the skill was employed. One of the teacher's important functions is to remind the student, on suitable occasions, that what he does in the science laboratory can have significance in other fields of thought.

A final word is given about the teaching of skills with maximum flexibility and generality. The categories in Dimension 3 are ordered beginning with those of most immediate sensory data and progressing toward those that are most abstract. It bears repeating that this is not intended to mean that their use must be invariantly in that order. Scientific investigation proceeds by many routes and is certainly very fluid as to the combinations and sequences of skills that may be employed in reaching a solution. Our cubical model is not intended to be a prescription of procedure; rather it is an organized way of holding the various categories in mind so that reasonable and informed choices among the categories can be made. In some laboratory experiences, with adequately mature students, one may begin with a deductively constructed model and then proceed to observation. Evaluation may precede interpretation of events in some experiments. It is very important to remember that flexibility in the use of science skills (on the part of the student) and flexibility in use of teaching methods (on the part of the teacher) will contribute to productive and creative laboratory experiences.

In this context of considering flexible use of science process skills, we proceed to a discussion of the kinds of scientific research methods that a student can be helped to understand and use.

Purpose 3: To help the student appreciate and in part emulate the role of the scientist.

Contemporary methods of science investigation vary so widely that it is impossible to make a universal and exhaustive categorization of them. However, for the purposes of science teaching, it is useful to have in mind some of the methods employed by scientists in research. Those to be cited here are (1) methods that I have observed in science practice and (2) methods that are compatible with the model of science presented in this book.

Let me state at the outset that I do not believe that all science is theory-based or of the controlled experimental kind. There are some strong voices in the field of science analysis who seem to purvey the notion that the only valid interpretations of modern science are those of theory-based research using theory-derived hypotheses to be tested. I have observed and participated in scientific research where the methodologies were quite divergent from this narrow view. The purpose of the classification of science research presented here is to provide the teacher with a way of thinking about diverse roles that scientists perform when engaged in scientific investigation. Such a classification can be useful in thinking about the various kinds of science laboratory experiences that students can and should have.

I identify seven kinds of science investigations: (1) direct observation, (2) multiple trials, (3) theory-based observation, (4) theory-based trials, (5) theory-based controlled experimentaiton, (6) theory construction, and (7) theory abstract analysis. The sequence is roughly organized in a series beginning with those science methods that are most dependent on immediate sensory data acquisition and involve the least amount of abstract theory and proceeding to those methods requiring increasing use of theory to guide research.

Direct observation here means descriptive science. The main thrust of such a method is to apply keen analytical skills to data-gathering as a means of generating an orderly explanation of the kinds of things one has observed. The ultimate product of this kind of research is a systematic way of organizing percepts, but not an explanation of phenomena that is so comprehensive and generative of new data as to qualify as a theory. The final product of direct observation is a more orderly assemblage of categories of sensory experience. Much of the work of early taxonomists, meteorologists, and anatomists was of this kind. It continues to be a significant preparatory step in much modern research. As an example, a student who observes the motion of a spider and develops a classification scheme for the various movements he has observed is conducting scientific investigation through direct observation.

Multiple-trials research is a kind of investigation that employs some manipulation of the environment to gain data. In this case, the scientist systematically tries one variation after another (multiple trials) to determine if there is an effect on another variable. There need be no theory guiding the kinds of variations employed; rather, there is an informed expectation that a certain kind of variable should have a certain effect on another specified variable. This kind of research has been used with much success in pharmacology. Antibiotics have been identified and isolated by screening techniques of the multiple-trial kind. Various organisms that are suspected of having bacterial inhibitory functions – either because of their observed activity in the natural environment or because they belong to a certain class of organisms – are cultured and applied to bacterial growth as a means of determining which if any will cause inhibition. When a potentially active one is identified from these trials, then some further growth and refinement in activity through selective breeding is attempted.

The lack of theory in multiple-trials research does not mean that it is irrational. The kinds of models that guide such research, though relatively rudimentary, consist of informed categorizations of likely variables to be tested and tried in turn. There may be some science curriculum writers who *a priori* reject this as a method of science; yet the fact remains that it continues to be used by legitimate scientists with productive results. Students can benefit by understanding that this kind of research is part of the enterprise we call science.

Theory-based observation is the use of a theoretical explanation of phenomena to guide the selection of phenomena to be observed and to make predictions about events that are likely to occur so that proper care can be taken during observation to determine if they do indeed occur. A *theory* (in my use of the term) is a thoroughgoing abstract explanation of the relationships among variables. It is a logically derived set of relationships that explain natural events. Its complexity of organization, generality of representation, and capability of generating predictions are far greater than in a simple model or hypothesis or general rationale or fragmentary set of explanations. This means that in theory-based observation the observed events are selected on the basis of a comprehensive explanatory scheme. This is to be contrasted with direct observation that is not initiated by a theoretical view.

As an example of theory-based observation, scientists have shown great interest in observing solar eclipses with the aim of testing the prediction – in Einstein's theory of relativity – that curvature of light rays is possible when light passes in the vicinity of a star. Such observation is initiated on more general and comprehensive grounds than merely to determine how light behaves. Since the objective of the research is generated from theory, the findings of the observer can do more than tell us how light behaves; indeed they can provide further evidence to support or in part negate the theoretical framework that gave rise to the observation.

Theory-based trials are akin to theory-based observation in that the various trials employed in research are either suggested by a theory or are used to further test certain theoretical assumptions. Some genetics research consists of theory-based trials. Based on Mendelian genetic theory, one may decide to try to develop a better quality of corn. To pursue this end, it is necessary to try various crosses among plants to eventually achieve the desired offspring. The difference between this method and "simple" multiple-trial research is that here at each trial the data are analyzed in terms of the guiding theory and the next trial is appropriately organized in relation to theory. Some modern particle physics research can also be classified in this category. Thus, to the extent that some theory about atomic particles exists, particle accelerators can be used to segregate various kinds of particles through multiple attempts and eventually it is possible that data about particle interaction can be obtained to further extend the theory or provide evidence in support of or contradiction to the theory.

Theory-based controlled experimentation is a method of testing hypotheses derived from theory using controlled experimental situations. The hypothesis is a statement of relationship between two or more variables as derived from a theoretical explanation and subjected to verification using an experimental approach where one or more control groups are compared to the experimental treatment groups. The experiment is usually contrived in such a way that the experimental variable is the only relevant factor varied among the groups. If some effect of the experimental variable is observed in the treatment group as predicted when compared to the control group, then evidence has been obtained that the experimental variable has caused the effect.

Consider the simple example of a theory-based prediction that X causes Y. To gain evidence that this relationship holds in the natural environment, one can perform a controlled experiment. A sample of test subjects or objects is drawn and randomly assigned to an experimental and a control group. All characteristics of the two groups so formed are the same in so far as random assignment will allow and supposing that the number is sufficiently large. Now, one group is treated with X and the other group is not. If Y occurs in the treated group, but not in the control group, then we have obtained some evidence that X is a causal factor for Y. This is of course one simplified example of a controlled experiment. Consider a more concrete example. In biology, modern genetic theory assumes that characteristics of organisms are transmitted from parent to offspring as inheritance units called genes. Moreover, molecular biology has further refined the theory to include the concept of molecular DNA segments as equivalent to genes. The stability of the molecular structure of a gene is essential to its reproducible effect. If a "gene molucule" becomes reorganized or otherwise severely perturbed, it changes its expression, producing a different phenotype than the original gene. The gene is said to have mutated. This theory yields a logical prediction. We know heat increases molecular entropy; therefore, if we expose an organism to elevated temperatures it should undergo increased gene disorganization and hence produce greater numbers of mutant offspring than would be obtained at more moderate temperatures. The hypothesis would be: The rate of gene mutation in an otherwise normal population of organisms is directly related to the environmental temperature.

To test this theory-based hypothesis, one can perform a simple contolled experiment. A sample of normal breeding Drosophila flies is randomly divided into two. Each sample by this technique should contain approximately the same ratio of males to females if the randomization was carefully performed and the initial sample was large enough. Now one sample is grown at moderate temperatures, and the other sample is subjected to sublethal elevated temperatures. The number of aberrant offspring in each sample is examined after the incubation. If the heat-treated sample repeatedly shows greater numbers of aberrant offspring, evidence has been obtained to support the hypothesis. The theory that yielded the hypothesis is also strengthened.

Many forms of controlled experimentation are used. The simple models cited above are not intended to be exhaustive or exemplary of the best; they are offered to help clarify the concept. We must recognize moreover that theory-based controlled experimentation is not the only type of controlled experimentation. It is the most elegant form. However, some important discoveries have been made using non-theory-based controlled experimentation. Instead of a well-developed theory, one may have either a general rationale that yields a hypothesis or simply an intuition, based on cumulative but not formalized interpretations, that a certain relationship exists in the environment. I have not chosen to separate this category, but it should be recognized as a legitimate, though less elegant, form of experimentation.

Theory construction may involve the gathering and interpretation of sensory data, but the major thrust is to build a theoretical explanation of phenomena. Much of scientific effort eventually culminates in theory production, and therefore it may be somewhat spurious to segregate this method. However, there are clear and notable examples where a person has set forth to generate comprehensive and general explanatory schemes as the purpose of his research, or where in any case the culmination of his research efforts was a complex theoretical explanation. Charles Darwin was such a person. It is obvious that Darwin did not begin *de novo*. He was well tutored in prevailing theories of the origin of life, and his theory was a logical outgrowth of some existing ones. But it was clearly so much more sophisticated and comprehensive that it qualified as a novel construction. Much of Darwin's theory was based on interpretations of sensory data gathered by Darwin himself. Some theory construction of course depends much less on immediate sensory data interpretation. Those theories that are derived from highly abstract interpretations, several times removed from the initial sensory data source, are produced by what I call theory abstract analysis.

Einstein's relativity theory is an example of *theory abstract analysis*. Scientists who construct or refine theory by interpreting and manipulating symbolic representations (such as mathematical models), rather than using direct sensory data, are doing abstract theory analysis. As we have emphasized in Chapter Two, however, some of the initiatory events in any scientific theory construction must be logically linked to sensory data reception, and preferably the analysis should yield deductions that can be subjected to empirical verification.

The foregoing sequence of seven representative kinds of science research has a correspondence to the sequence of process categories in Dimension 3 of our model. This correlation is shown graphically in Chart 1. As one proceeds through the sequence of seven research methods, it is clear that more and more of the advanced process categories in Dimension 3 are being included. It is not possible to make hard and fast separations among the categories included in each scientific method, but Chart 1 does indicate which inquiry processes are most typically involved in each kind of research design.

Direct observation and simple *multiple-trials* research are most directly related to the observation, integration, and interpretation categories of Dimension 3, with least emphasis on interpretation in both cases. *Theory-based observation* and *theory-based trials* include observation, integration, interpretation, and prediction as their most representative processes; prediction enters here because a theory is typically more generative of predictions than are less complex and comprehensive models or rationales. Some evaluation may also be involved in theory-based observation and theory-based trials, but it is not typically essential. We do not chart the higher process categories, reconstruction and construction, as belonging to any of the three "theory-based" research methods because the theory guiding such research is used mainly, or at least initially, as a model to allow systematic data-gathering; the main objective of the research may not be to create a new or revised theory. *Theory-based controlled experimentation* can include all process categories from observation through evaluation; and indeed evaluation becomes a major process for this kind of research, in which hypothesis-testing is a critical step. Theory-based experimentation sometimes also includes reconstruction, but not typically. *Theory construction* and *theory abstract analysis* can include all of the processes of Dimension 3, but reconstruction and construction have the most obvious relevance for these methods. Theory abstract analysis in its most abstract form may not involve observation at all, since much symbolic theory-building, particularly in some fields of theoretical physics, does not depend on immediate sensory data but uses previously gained constructs to form or elaborate a theory.

CHART 1

Science Process Categories Related to Research Methods

Note: The varying width of each horizontal bar indicates the relative significance of the science processes in a particular research method.

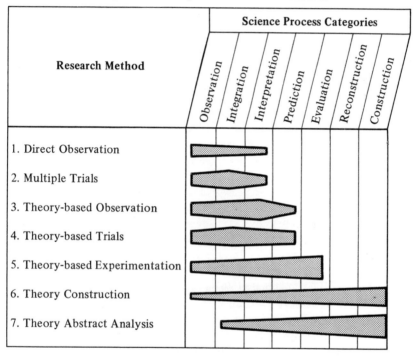

The correlation presented in Chart 1, between science methods and process categories, can serve as a practical guide in designing curriculum experiences to match the particular abilities of students and to plan for growth through approximations toward more complex experiences. Research Methods 1 and 2 require less abstract thought to be implemented than do the remaining methods. Direct observation and multiple-trial laboratory experiences can be used at an appropriate difficulty level for less advanced students in science. It should be clear that although these research methods do not require as much abstract conceptualization as some of the others, they can become very elegant as to the clever use of interpretation and the design of procedure for data gathering. Therefore careful analysis by the teacher or curriculum designer is needed to properly select experiences in these categories to match student abilities.

Students who are sufficiently mature can be introduced to Methods 3 and 4. These are appropriate as an intermediate step in progress toward more abstract methods of research. When students are still more advanced in science understandings, they can be introduced to Method 5 (theory-based controlled experimentation). Very creative students with developed capacity for abstract thought can eventually use Methods 6 and 7 in their most elegant forms — but only as applied to data simple enough to match the students' current store of science knowledge. It is obviously absurd to expect high school students to elaborate quantum theory. It is more reasonable to assume that they can identify theoretical constructs concerning such things as the melting of solids, light propagation using simulation models, and so forth.

A final word to refine this discussion on matching laboratory experiences to student maturation. First, it should be kept clearly in mind that laboratory experiences should not be used merely to maintain student competence at a *status quo* level. Students should be encouraged to develop skills in more and more abstract forms of science methods in so far as they are capable. This requires careful planning to create a curriculum sequence that will challenge each student to grow to a level commensurate with his intellectual ability. Secondly, although the seven research methods cited here have been ordered to reflect various levels of abstractness, it is possible to design experiences at all levels but one (Method 7) that are simple enough in kind of knowledge required, and sufficiently concrete in their early stages of use, that most average students can at least come to appreciate the role of the scientist in performing these methods. Some students may develop explanations sufficiently sophisticated to qualify as "microtheories" — simple models of animal behavior, of wave motion in Slinky springs, or of chemical reactions, for example. Obviously only a few students of particular intellectual ability and creativity will be able to develop more elegant explanations.

With average students, it is wise in my view to concentrate intensively on the lower-level methods. It is by very thorough grounding in these that students can be helped to appreciate the orderliness and tentative nature of scientific thought and to develop precision and rationality in information-gathering. Attempts to coerce students into emulating the most elegant accomplishments of scientists will usually fail — and, in failing, will subvert the vaild purposes of science instruction. Some laboratory experiences exhibiting the various kinds of research methods are described in Chapter Five, and their implications for teaching are discussed.

Purpose 4: To help the student grow both in appreciation of the orderliness of scientific knowledge and also in understanding the tentative nature of scientific theories and models.

In recent years, school curricula have placed increasing emphasis on instruction in "critical thinking" and the "processes of science" (as distinguished from theories and bodies of knowledge). Clearly these are emphases, sometimes neglected in the past, that are appropriate to science teaching. However, we must not lose sight of the importance of helping young people understand that man can order and explain experience so as to interpret reality coherently, and that the theories and bodies of knowledge created by scientists over the centuries, systematically observing and interpreting natural phenomena, represent one of man's most impressive and comprehensive attempts to understand himself and his environment. We cannot behave responsibly as mature individuals if our only ability is to criticize – if we do not have enough organized knowledge to appreciate and use the contributions of the past in the making of new discoveries and vital decisions. Science instruction should foster observation, integration, and interpretation skills aimed at helping students believe that man is an orderly and responsible being. Science is a dynamic complex of acitivity, combining critical analysis with knowledge accumulation and orderly explanation. To emphasize one aspect to the near exclusion of the other is not truly representative of science.

Science is also dynamic in the sense that its theories are not accepted as fixed. It is advisable to provide learning experiences – as of model construction and reconstruction – that will lead students to realize the limited nature of their own constructs and to revise their interpretations of reality as they gather additional, sometimes contradictory, data. An alternative to such direct experience is discussion periods in which various historical examples of theory construction and verification are analyzed. In physics, for example, the theory of the atom has undergone massive revision from time to time as new data were collected and applied toward construction of a more comprehensive explanation. The contributions of people such as Rutherford (who added to our understanding of atomic charge distribution) and Bohr (who devised the solar system model) and Schroedinger (who devised a highly sophisticated mathematical view of the atom) illustrate the dynamic changes that science theory development has undergone. Classroom discussions about the theoretical views prevailing at the time when a certain new conceptualization came forth, and about some of the science-historical events antecedent to that new conceptualization, can be included to make this an interesting conversation. Where appropriate, comparisons can be made between the earlier theoretical revisions and modern advances in theory revision and construction. It should be helpful for students to consider rationally the evidence that necessitated or allowed the new formulations to come forth.

Preparing Students
for Laboratory Activities

We have considered some purposes of laboratory instruction and their implementation. We consider next the preparation of students for laboratory experiences. In Chapter Three, I introduced the idea of *schemas* and briefly explained how these consistent ways of organizing experience can influence what one attends to and how one interprets sensory data. In this chapter we have discussed among other things the kinds of methods a teacher can use in teaching scientific skills and introduced several categories of scientific methods to elaborate the basic cubical model. These several ideas are further amplified here, by combining them in a discussion of ways to prepare students for laboratory experiences.

The amount and kind of preparation a student needs for a laboratory experience will depend upon his prior general knowledge in science education, his ability to think systematically and creatively without assistance, and/or his ability to gather information as needed during a laboratory experience.

In addition to the student's characteristics, the kinds of teaching methods that can be used in the laboratory are also relevant to the amount and·kind of advance preparation. If laboratory work is to be presented at an average difficulty level for most of the students, and if the teacher plans to use a good deal of prescription as opposed to free-selection techniques (see p. 70), then there need not be a great deal of student preparation for the laboratory experience other than the requisite knowledge to begin the experience. The teacher will be available to give guidance through much of the experience. If the laboratory experience is to be above average in difficulty, and if the teacher plans to use a good many free-selection techniques of teaching, then more attention may be warranted in preparing the students for laboratory experiences. The preparation should moreover be consonant with the purpose of the laboratory experience and the teaching technique to be employed. For example, if one plans to use much free selection in a laboratory experience that requires some development of prediction and evaluation skills, it would not be appropriate to give so much detailed information ahead of time that the student's behavior is largely predetermined by the preliminary instructions and information given.

In preparing a student for laboratory experiences, we can distinguish between (1) giving knowledge and (2) describing skills. To *give knowledge* means increasing the students' store of factual, conceptual, and abstract information. Knowledge thus includes memorized information about the findings of science apart from the methodologies used. To *explain skills* is to present information about the methodologies or processes of science. These are the particular intellectual (logical) skills and laboratory manipulative skills used by a scientist. It should be clear (as we have pointed out in Chapter One) that in practice it is not possible to separate the two (knowledge gained and skills used), since they have a dynamic interactive influence on one another. However, in preparatory teaching, it is logically convenient to make a distinction so that we can more precisely determine what information will be supplied in advance of a laboratory experience.

In addition to the distinction between background knowledge and skill preparation, we can use three ways of presenting either kind of information, in terms of *schema preparation.* The three ways are (1) schema mobilization, (2) schema reformulation, and (3) schema formation. Recall that a *schema* is an organized way of viewing experience which includes one's attitudes toward certain kinds of experience and the cumulated knowledge relevant to these experiences. Most if not all students in high school and college will be sufficiently experienced to have well-formulated schemas. This is particularly true for general life activities and, depending on prior general science training, will be more or less developed relative to science experiences. Obviously, students who have had considerable science training in elementary school and the middle grades will have more organized knowledge and attitudes to bring to bear on a situation than less fully educated students. We consider now the three ways for using such organized ways of viewing experience as preparation for a laboratory experience.

Schema mobilization means that the teacher intends to provide little new information to students in advance of the laboratory experience; rather, the teacher intends to help the student recall and organize prior-gained information so as to prepare the student to deal with new problem situations.

Schema mobilization is facilitated through a discussion technique. In this approach, the teacher raises certain problem questions in the subject matter area to be used in the forthcoming laboratory experience. He then helps the students call forth prior-gained information and skill descriptions that can be useful in such problem areas. The discussion need not be a long one, and it should certainly be centered in some interesting problem. Suppose a teacher plans to have a laboratory on isolation of fungi that might produce antibiotic substances. One of the skills he wants the students to use is proper bacteriological culture and transfer techniques. He may know that the students were introduced to such practice in junior high school. However, the students also learned the description of a fungus life cycle. To prepare the students for their lab experience, he would like to have the students mobilize these schemas of bacteriological culture skill and knowledge of fungal growth.

The discussion can begin with the problem of the medical researcher who would like to isolate a fungus from among many he has that will retard bacteriological growth. What kinds of information can he use about fungal growth, and how can he proceed? Through dialogue with the students, the teacher should get them to recall the life cycle of fungi and to state the kinds of culture techniques that can be used to grow both fungi and bacteria. The teacher may then state that in the next laboratory period students will attempt such isolation and that at that time they will be able to figure out a strategy for the needed procedures, based on their present discussion. The laboratory can then be arranged with as much student autonomy as the teacher feels is warranted on the basis of the discussions.

Note that in schema mobilization the teacher does not add appreciably to the students' fund of information, but merely helps the students mobilize relevant information that they already possess.

Consider next *schema reformulation.* Here the teacher either adds to the students' fund of information and/or induces the students to reorganize their thinking about prior experiences in an area relevant to a forthcoming laboratory experience. Consider again the fungus experiment cited above. Schema reformulation techniques might proceed as follows. The teacher first states the problem as previously given, but then gives the students some new conceptual information on the isolation of penicillin from the fungus *Penicillium.* As he does this, he may ask the students to explain from their prior experiences some of the methods in bacterial culture and fungal growth that clarify key points. He may also ask the students to consider how they might reformulate some of their prior-gained methodology in light of the new information presented. Here it is clear that the teacher has added new information to a mobilized schema and has also suggested, or induced the students to suggest, ways of using methods in a new approach. This, then, is an instance of schema reformulation.

Schema formation is a much more extensive operation. It presumes that the students have little prior information in the area under consideration and that some preliminary structuring is needed before the students enter the laboratory experience. Therefore, the teacher or assigned student discussants will present some main organizing ideas that are new to the students. They are some relevant conceptual explanations and generalized mental sets about the way a laboratory experience is to be approached. As an example, using once again the fungus experiment, the teacher with key student assistance would build up ideas about Fleming's work in penicillin isolation, about the relevant kinds of techniques to be used, and about some general problems in this kind of research. Most of the information is new and therefore constitutes schema formation.

A stimulating and creative teacher will plan to use various kinds of preparatory experiences with students from time to time. To use schema mobilization or reformulation requires careful analysis of the students' prior educational history to identify the kinds of knowledge and skills they have previously encountered.

One other refining point needs to be made in connection with schema preparation. The teacher can help the students identify salient points in a schema that might be particularly crucial in a forthcoming laboratory experience. Certain bits of knowledge or particular points of skill application critical to the lab experience can be highlighted during the preparatory phase. I call the process of highlighting certain key points of information *image labeling*. Thus a schema is, as it were, an image of experience which we are in some way preparing. If we emphasize certain points by illustrating them in several different ways, or by including in the preparatory discussion a particularly poignant example of their use in the history of science, we have given potency to these ideas. The image becomes *labeled* at these points and therefore more ready for recall and application in a subsequent laboratory period. This requires careful thought on the part of the teacher as to what ideas may be very crucial to an experiment and what therefore should be highlighted in an informal way during the preparatory presentation. Consider the example, as in a previous discussion, where a teacher is preparing students for a fungus culture lab. At one point the procedure for bacteriological culture technique is elicited. To image-label this procedure, the teacher can show a film loop on this topic. Thus it has been presented in two different modes, increasing its saliency and availability in memory.

Image labeling is used in highlighting ideas as a mental function. To be complete in our discussion we must also consider *object labeling*. This is simply presenting an object or set of objects wherein certain parts have been given saliency by use of color, name labels, or pointers. Consider the simple case where the teacher of chemistry wants students to attend particularly to the properties of halogens. This can be accomplished by presenting a periodic table where the halogen family is tinted red on the chart. The students are then attracted to attend to them. Some preserved biological specimens are available with color-latex-injected organs to make them more salient upon dissection. For example, the blood arteries are injected with red and the veins are injected with blue. This makes the two circulatory system components more salient and aids the student in visual tracking (perceiving the overall plan of each system). The concept of object labeling is very simple, and is introduced here merely to complete the discussion of *labeling*. Image labeling concerns highlighting certain ideas in memory, and in a similar way object labeling is the highlighting of objects to be observed.

Schema preparation, including image-labeling techniques, can be used to orient students to any number of skills discussed previously in this book. For example, if there is to be use of certain science process skills (Dimension 3) in a laboratory session, some preliminary discussion of them will help prepare the students for the experience. The teacher may find that students will elect to talk about issues related to orientation toward the natural and social environment (Dimension 1). In some cases, schema preparation is needed to enhance the students' readiness and susceptibility to ordering and systematically categorizing data. Some kinds of topics that can be used here are great historical scientific explorations that resulted in more systematic knowledge of phenomena – including brilliant conceptualizations of the past (such as Mendeleev's development of the periodic table) and events in the continuing story of modern physics (discovery and classification of the atomic particles). The teacher may also want to enhance value system schemas by discussing the criteria scientists use in deciding what projects are worthy of exploration, what kinds of methods are most appropriate to a problem under investigation, and what standards of excellence in evaluating results are to be used.

Schema preparation and image labeling can be accomplished by other methods than classroom discussion. For example, a written communication can be used; it will have to be composed to accomplish one of the three kinds of schema-preparation methods: mobilization, reformulation, or formation. It is also possible to use a preliminary laboratory experience to act as a schema-preparation step; the experience is organized to arouse student awareness of concepts and methods he will need in a subsequent laboratory experience. As an example, a physical science teacher may want to form a schema about wave interaction in preparation for a laboratory on diffraction using light sources. The preparatory lab could employ ripple tanks, to help the students form a schema of wave interference and to develop analytical skills in understanding the mechanisms of constructive and destructive wave interference.

From the foregoing discussion it should be clear that schema preparation can be very subtle, as in schema mobilization and reformulation, or more direct, as in schema formation. Some such preparatory step will be helpful for many students before entering a laboratory experience. Due attention should be given to this function of teaching, to moderate the kind and amount of preparatory experience in accordance with students' need and with the difficulty of the forthcoming laboratory experience. Schema preparation should occur as soon as possible before the laboratory experience – ideally, not more then one or two days prior to the experience. Otherwise the organizing effect of the schema preparation may decay.

Psychological Functions
of Science Laboratory Learning

Laboratory experiences can serve several functions, and in some cases should serve two or more functions simultaneously. Earlier in this chapter we discussed some purposes of the laboratory and issues relevant to their implementation. That discussion led us to consider areas of laboratory experience that relate to developing knowledge and skills relevant to the new perspective established in this book. Those ideas are equally pertinent in this section of the chapter, but here we turn our attention to some general psychological functions of the laboratory experience.

In addition to increasing the student's knowledge store, his repertoire of skills, his expanded awareness of the role of the scientist, and certain enhanced ways of viewing the environment aesthetically and rationally, the laboratory experience can also aid efficient *organization and storage of new information* in memory. In Chapter Two, we pointed out Suchman's view that students acquire greater meaning for and permanence of learned material when they organize and explain experience with a certain degree of autonomy. Helping the student to mobilize his own cognitive resources and to discover meaning in his own experience is one function of an open-ended laboratory experience of the kind we have called free selection.

Laboratory learning can aid also in *consolidation of imformation* presented in close temporal proximity to it — i.e., knowledge acquired no more than a day or two in advance of the laboratory. To pursue this topic we will need to discuss the concept *consolidation* as psychologists understand it.

When we perceive something, let us say by hearing a verbal communication, the stimulus is briefly sustained as a mental representation in what is known as echoic memory. The verbal information, in so far as it has been faithfully perceived, is sustained for this brief period of time in the form that it was transmitted. Thereafter, the stimulus material is transferred into a temporary storage called short-term memory. This phase of the information-processing is usually rather fluid in organization. The acquired information undergoes some recoding. For example, the information may be organized into large units; long strings of information can be lumped together, thereby reducing its complexity and magnitude and rendering it more susceptible to being committed to long-term storage. Short-term storage, as the name implies, is a temporary short period of storage. It may last from several minutes up to several hours. During this period, the information received is reordered, reorganized, and related to prior-gained information as a way of making it suitable for storage in long-term memory. Long-term memory is that part of memory that holds information for as long as it can be recalled. It is the final repository, as it were, after short-term memory. The effectiveness of long-term memory storage is determined in part by the amount and type of organization the person imposes upon information during the short-term memory phase. This process of short-term memory organization is called *consolidation*. It is the process whereby newly acquired information is interrelated and associated with prior-gained information. In other terms now familiar to us, it is a mechanism of integration – of relating newly acquired information to previously formed schemas, thus making it more durable and meaningful in long-term memory. Some of the environmental factors that influence consolidation are the amount of time given for active processing of received information before new information is presented. If a rapid sequence of information is presented, the short-term memory consolidation processes can become disorganized by an overload of information that cannot be adequately organized for submission to long-term memory. Conversely, adequate cueing or repetitive presentation of information after initial presentation can aid consolidation. Thus, if information is presented and if subsequently, after a short interval of time (several hours to about one day), there is a repeated encounter with this information in a way that aids its organization and association with prior experience, then the consolidation and submission of the information to stable, long-term memory is enhanced. It is important, however, that the subsequent experience provide appropriate opportunities for the individual to be actively engaged with use of the information in an organized way that allows its meaningful association with prior experience.

The well-organized laboratory experience facilitates consolidation. It provides opportunities for students to become individually involved in organizing, reinterpreting, and integrating new information with old. If proper preparatory work has been done in establishing saliency of relevant schemas, and if the laboratory experience is an orderly process of investigating phenomena related to prior-presented information, consolidation will be aided and long-term storage enhanced. We must be clear in our understanding that even the most stable store of information in long-term memory will undergo transformation and perhaps decay so that it will not be as readily available at a later time as it was immediately after storage in long-term memory. However, to the extent that we can arrange for facilitation of consolidation as information is transmitted to long-term memory, the more available it will be and the more readily recovered upon subsequent review. Moreover, recall of information not immediately available under conditions of few cues may, when properly consolidated for long-term storage, be recalled in contexts (environments containing cues) similar to the learning situation. This is reminiscent of the discussion presented on transfer of training and is obviously related to it.

The main thrust of all of this discussion about consolidation is to remind the teacher that appropriately organized learning experiences requiring active student participation can enhance long-term memory of antecedent information. The laboratory experience is an excellent opportunity to realize this advantage. The use of integration as represented in Dimension 3 of our model is a way of facilitating meaningful association of newly acquired information with that already contained in organized memory.

This use suggests careful analysis of the curriculum sequence to arrange laboratory experience sufficiently soon after relevant information reception to aid in consolidation of that information, and also to extend the student's store of knowledge and skills, during the laboratory experience. The laboratory experience can serve several educational purposes simultaneously if sufficiently careful thought is given to optimal sequencing and organization of the experiences. It is important to keep in mind, however, that one can try to do too much in a laboratory period and thus overshoot the students' ability to handle the data. Therefore, whatever consolidation function goes on in a laboratory experience, it should be sufficiently compatible with new learnings offered that the student is not intellectually overburdened. Indeed, not every laboratory will serve as a consolidating experience for prior-gained information. The purposes of the laboratory experience can be simply to introduce new content and to teach science process skills.

In our discussion of the preparation of students for a laboratory experience, attention was given to the development of schemas and the role they perform in guiding student laboratory experience. In Chapter Three, schemas were defined and the role of the teacher in helping students achieve greater *schema differentiation and expansion* was discussed. This theme is further developed here.

The degree of intervention of the teacher in the laboratory experience will depend partly, as previously noted, on the kind of teaching method used. Within the context of method applied, the teacher should be alert to identify the mental organization that a student brings to bear on a situation and help him further refine it. At the most elementary level of functioning, the very awareness by the teacher that such antecedent organizing schemes exist is a useful aid. It calls to our attention that student behavior is mediated by preformed mental sets that govern what he observes and how he explains experience. Much of student behavior in the laboratory is not capricious but can be understood if one has a general understanding of the schemas the student brings to a problem situation. An alert teacher will attempt to identify the kinds of organizing schemes that a student possesses and thus be in a better position to assess and guide student growth.

In most cases, evidence for the kind of schema a student brings to a situation is at best fragmentary, and teachers need to be alert to detect cues that indicate the organized representations that a student has about experiences. I cannot offer an algorithm guaranteed to yield a full and complete understanding of a student's schema. I can, however, present some guidelines that will assist teachers in remaining alert to identification of the kind of organization a student brings to bear on a situation.

The *kinds of questions asked* by a student will often reveal the comprehensiveness and degree of differentiation of his schema. A question about a very fundamental aspect of a task may indicate a complete lack of comprehension of some basic information necessary for the task. Or it could mean that the student has a temporary remission of memory on this point. To determine which of these alternatives is correct, you can ask a few additional questions in the same area to probe for the student's knowledge. If he shows little further understanding, then there is good evidence that his knowledge is deficient in that general area and he will need assistance in gaining some basic information. But if after questioning he shows some comprehension of related information, he has probably suffered a mild attrition in his memory and a little assistance may be all that is required to bring his schema to adequate functioning level.

A second indicator is the *kinds of statements uttered* by a student. A statement of specific fact that is erroneous is the simplest kind of indicator. This may represent a minor misunderstanding of a particular fact. To determine whether there is a larger misunderstanding, one can once again ask a few general questions in the same area and thus determine the range of a student's accuracy in knowledge.

There is good evidence to show that students organize knowledge in a hierarchical manner, beginning with the most general information at the top of the hierarchy and progressing toward more specific information at the lower levels.[2] Therefore, in probing to determine a student's accuracy of knowledge in a given area, it is wise to begin with more general questions and proceed with more specific questions to determine where his knowledge is deficient. Care should be taken in the process of probing so as not to overly burden the student with questions. A carefully thought out sequence of one general question and two specific questions should usually be sufficient to assess a student's general schema in an area of knowledge. This process is called a *hierarchical knowledge probe*. As an example of hierarchical knowledge probing, suppose a student is observing a frog circulatory system and inappropriately points to a blood vessel and calls it an artery. To determine the level at which the student's understanding fails, the following three questions could be asked:

1. What kind of circulatory system does the frog have — open or closed?
2. What are the two kinds of vessels in the frog's closed circulatory system?
3. Do the veins or the arteries carry blood back to the heart?

2. B. J. F. Meyer and G. W. McConkie, "What Is Recalled after Hearing a Passage?" *Journal of Educational Psychology, 65* (1973), 109-117. For background information on theories of knowledge organization and learning, see O. R. Anderson, *Teaching Modern Ideas of Biology* (New York: Teachers College Press, 1972), pp. 40-88.

This should bring the teacher back to the specific level where the student made an error in vessel identification, and the correct identification should then follow logically from this line of discussion. As a teacher, you will require some practice in being prepared to perform such hierarchical probes, and a little conscious effort to use the technique on several occasions will facilitate more comfortable use of it in the future.

A third indicator of schema disorganization or distortion occurs when a student utters in succession two or more statements that are inappropriately juxtaposed. I will consider two kinds of cases. (1) An *illogical juxtaposition* occurs when two or more statements are presented together where one or more do not logically follow from the others. (2) An *erroneous juxtaposition* occurs when two or more statements are grouped together that are not related to one another given the present state of our knowledge. Clear distinction should be made between erroneous juxtapositions and novel or insightful combinations of information. The latter must not contain an inconsistency and must be a reasonable association of two ideas.

The following is an example of illogical juxtaposition: "Sodium chloride is an edible salt. Sodium fluoride is related to it and therefore must be edible also." This statement pair represents an overgeneralization and shows that the student is not sufficiently aware of the differences in reactivity of the two halogens chlorine and fluorine. This illogical juxtaposition may indicate that the student's schema is not sufficiently differentiated to allow him to recognize that although fluorine and chlorine are congeners in the periodic table, they are very different in their effects on biological systems, particularly in the anionic form. Another possible problem indicated by such an illogical juxtaposition is a mental set to overgeneralize in interpreting data.

An example of erroneous juxtaposition is: "This spider has four eyes. Is that common in most insects?" This indicates a lack of proper organized knowledge about classification of organisms. Spiders are not insects. They belong to the taxonomic class *Arachnida*, which is of course separate from the class *Insecta*. To determine the level of error in this erroneous juxtaposition, this series of hierarchical probing questions might be posed:

> To what phylum do the spiders belong?
> Are insects and spiders both in the same phylum?
> What are the different classes to which they belong in the phylum *Arthropoda?*

Such questioning not only reveals the level of a student's knowledge but also, through discussion about each point, increases the student's knowledge and precision in thinking about the knowledge he has gained.

A fourth indicator of schema disorganization is *misapplication of a skill or method.* A student who attempts to use a skill in an inappropriate context is showing evidence of confused purpose in using a skill. For example, a student may attempt to assess the amperage in a circuit by applying a voltmeter when there is no information available on the resistance of the circuit. Another example is the student who attempts to boost a DC voltage by inserting a transformer in the circuit. The misapplication of some abstract mental skills is also included here. An example is an attempt to apply statistical tests to data that have not been gathered in a way appropriate for the test. The student can be helped to correct his schema by identifying the salient characteristics of the skill and determining the conditions in which the skill is appropriately applied and those in which it is not.

A learning phenomenon that is closely related to schema development is *acquisition of a predisposition* — one of the psychological process orientations of Dimension 2. These orientations include attitudes about the presence of order in experience and about one's own ability to pursue orderly explanations. Since these attitudes are tacit ones, they cannot be instilled by verbal hortatory methods alone. A predisposition is developed as a stable way of approaching phenomena only after repeated positive experiences of a kind that promote the skill to be acquired. If you plan to help students develop an attitude of expecting to find orderly and rational explanations for phenomena, then they will need to have repeated exposure, over a long period of time, to activities leading to such outcomes. The more opportunities one has to practice orderly interpretations of experience and the more often one is appropriately and regularly rewarded for such behavior, the more likely it is that one's predisposition to find order in experience will increase and persist over time. Tacit understandings of many kinds develop with repeated exposure to situations with a constant characteristic to be internalized.

Orderly thinking is not something automatically imposed upon us by the environment. The environment is rich in sources of repeated and minimally organized stimuli that facilitate human recognition of orderly patterns. But many modern psychologists and philosophers point out that the kinds of groupings and general patterned organizations man finds in nature are those that he creates and imposes on sensory experience. Many of us who have had the advantage of early and continuous educational experiences in orderly interpretation of phenomena may assume that this comes as "second nature" or by way of "common sense" to all individuals. We have had so much training in orderly representation of phenomena, from our preverbal childhood upward through the years of our maturation, that we accept it as innate. But some young people, not having had the benefit of such pervasive education in viewing experience rationally and in an orderly manner, desperately need help in developing this predisposition if they are to function productively in society. Laboratory science experiences should be planned on a recurring basis so as to allow and indeed facilitate construction of orderly explanations of experience. Some examples of such laboratory experiences are presented in Chapter Five.

Group Processes
in Laboratory Teaching

The organization of laboratory experiences, considered in foregoing sections of this chapter from several points of view, must now be considered in an additional context — that of student involvement. There are four traditional ways that students have been involved in laboratory experiences: (1) as observers of a teacher demonstration, (2) as participant observers of a demonstration, (3) as active individual participants, and (4) as active group participants.

Teacher demonstration with student observation allows the least amount of student involvement. This method should be used sparingly, especially with students who are capable of engaging in active laboratory experience. A basic assumption in this book has been that a primary value of the laboratory is its unique capacity to allow students to experience critical, creative, and precise analyses of phenomena in a first-hand way. Too much pure teacher demonstration denies the student the opportunity to develop autonomy in dealing with intellectual problems and to acquire a sense of aesthetic response to intellectual analysis of the environment. Teacher demonstration is a reasonable approach when an exhibition of a particularly elegant or expensive procedure is desired that simply cannot be made available to students individually or in small groups. For example, the teacher may want to demonstrate an electrocardiograph recording obtained on an oscilloscope. The sensitivity of the skill required and/or the expense of the equipment warrants this if it is consistent with the overall objectives of the curriculum.

In general, if teacher demonstration is to be used, then student participant observation, rather than passive observation, is clearly the method of choice. The students should participate with the teacher at least intellectually. Discussion can include diverse related topics derived from our three-dimensional model, including matters of significance as to current science affairs, the impact on society of the findings of science investigations of the kind in question, and so forth. In the case of student participant observation of an electrocardiogram demonstration, discussion can focus on the medical significance of the method, the physiological and anatomical functions that can be inferred from the EKG trace, and the limitations of the evidence as obtained with the instrument.

Various forms of teacher demonstration are effectively used as schema formation techniques and are certainly efficient ways of preparing students to pursue investigations on their own using the instruments and methods demonstrated by the teacher. As mentioned before, such preparatory demonstrations should occur not long before the subsequent laboratory experience but also probably not immediately before the lab unless the procedure to be used is complex and requires short-term memory to facilitate its application. Most preparatory work for a demonstration can come a day or two in advance of a laboratory session.

Active individual participation is a common way of engaging student activity in laboratory experiences. The student is expected to work more or less autonomously, depending on the amount of freedom planned within the teaching technique. This can range, as previously noted, from teacher prescription through student free selection of procedures to be used. Individual student participation should be used often enough to foster students' confidence in their ability to successfully broach intellectual problem situations. Some laboratory experiences may include initial individual participation followed by group analysis of the combined findings of individuals. This allows students to contribute individually and as members of a group in problem solving.

Group participation is a technique that is gaining prominence in science laboratory teaching. Problems selected for investigation are those that allow small groups to collectively plan an investigation, so far as that is feasible, and to gather data for interpretation. Group participation has the advantage of allowing students to develop problem-solving skills in a social context. This learning experience involves not only acquisition of scientific skills, but also skills in dealing effectively with social group interaction. The limited scope of this book will allow only a general discussion of small-group psychological processes. The information that will be offered here is intended to alert you to certain basic group processes that can either facilitate or retard effective small-group function. These clearly constitute an aspect of one of the categories in Dimension 1 of our model — orientation toward the social environment.

The concepts to be discussed are (1) group goal identification, (2) group norms, (3) group cohesiveness, and (4) group member roles. This sequence of topics is ordered from the most general group functions to those related to individual group member functions. The discussion will be organized around the general principles that interrelate these concepts. To orient the reader for that discussion, each of the concepts is first defined.

Group goal identification is a process of determining the purpose of a group. It includes a statement of the ultimate thing the group wants to accomplish as a result of its interaction. In a science laboratory group investigation, the group goal could be, for example, to construct a model of crystal structure bassd on inferential data. In addition to the substantive goal (the kind just cited), there can also be goals related to group processes. One of the purposes of the group members may be to identify common areas of substantive interest that will allow them to work more effectively with one another. In most cases, the substantive goal will be the predominant kind demanding attention of the group. *Group norms* are the acceptable ways that members of a group are expected to behave. *Group cohesiveness* is the degree of solidarity and mutual support shared among group members. The more cohesive a group is, the more the members work in a concerted and amiable way toward a goal. *Group member role* is a function of an individual in the group. It is all of those activities that an individual performs in his function as a group member. One role, for example, is that of group leader — one who tends to organize and regulate group action; this is usually a dominant individual or one with particular ability to help the group achieve its goal. Another role obviously is that of group participant — any member who acts as a contributor to realization of the group goal.

The dynamics of group activity will now be discussed in the context of these four concepts. We must recognize that this is a limited view of the many complexities that groups exhibit.

Group goal identification is an essential step toward effective group functioning. Unless a group has and clearly understands a general substantive goal that is acceptable to most, if not all, members, then there is likely to be continuous confusion and possibly conflict in that group. One way to make sure a group goal is clear-cut is to have it supplied by the teacher. This is an efficient way of getting a group mobilized to begin active work in so far as they are adequately committed to the goal and interested in it. If proper care has been taken in preparing the students for the experience through methods of schema development and so forth, then a prescribed goal is likely to be a successful guide for the group. In some ways however, it is preferable for a group to identify its own goal(s) in so far as is feasible. There is in general likely to be greater commitment to group-identified goals than to teacher-prescribed goals. Yet often a good advance discussion about the significance of a prescribed goal can sufficiently mobilize student interest to make it suitable. By a good discussion I mean a dialogue between teacher and students that allows points of students' natural interest in the goal to emerge.

In some cases, once a group has begun operation, members may establish subgoals that allow them to progress step-wise toward the ultimate goal. Each subgoal represents a reasonable partial accomplishment; taken collectively all the subgoals eventually summate to yield the final goal. Some care needs to be taken in identifying reasonable subgoals that are consonant with the ultimate objective to be achieved.

Group member roles are significant in determining the ease and rapidity with which goals are identified. If the group has developed an agreeable role structure where each member works amiably with the others, goal selection and/or progress toward a selected goal will be facilitated. On some occasions you will notice groups that do not seem to be making organized progress. Careful attention to what is being said may indicate the problem source. In some cases, particularly in newly formed groups, two or more students may compete for leadership positions. This competition can evidence itself in direct challenges as to who will be the leader, or it may appear in more subtle ways through indirect challenges. An indirect challenge can take the form of presenting one's abilities and declaring one's accomplishments. Such subtle bantering may not be simple peer competition, but can be a result of the group composition. The teacher can facilitate passage through this critical stage of role identification by assisting the group to rationally and sometimes democratically elect a leader and to assign other competent individuals to roles that match their abilities. This procedure allows the group to get on with the problem of goal completion and, if performed maturely, results in everyone being satisfied with his contribution and position. A group leader need not be an overseer, and probably *should* not assume this role, for best functioning of most groups. The leader should be a coordinator of group findings and moderator of discussion, in addition to contributing whatever other abilities and knowledge he may possess relevant to the group objective.

If group members are satisfied with their roles, and if a mutually reasonable if not fully desirable goal has been identified, group cohesiveness will usually increase. The cohesiveness of a group also depends on the degree to which a group has developed behavioral norms that are reasonably widely shared among the group members. Groups that have successfully identified member roles and started working toward a goal may still suffer conflict due to incompatible norms of behavior. The teacher can assist students to find mutually agreeable behavior expectations. You will need to listen carefully to the kind of dialogue that dominates a group in conflict to determine the apparent sources of behavior disagreement. If these are tactfully pointed out, sometimes mature group members can bring peer pressure to bear on those who are not contributing substantially to progress. When there is genuine intellectual disagreement that produces rational dialogue, this is to be encouraged. Indeed, intellectual nonconformity is probably a healthy sign, not to be suppressed. However, a group may need help from the teacher in dealing rationally rather than emotionally with divergent individuals.

Keeping in mind these four Dimension 1 social-orientation concerns — goals selection, group cohesiveness, norm development, and compatible role identification — the teacher can aid students in developing amiable and intellectually productive progress toward project completion. Aspects of the other category in Dimension 1 are discussed in our next section.

Aesthetic Values
and Functional Significance

When science is taught as a human enterprise, Dimension 1 must be attended to in conjunction with Dimension 3. To consider the issues generated by Dimension 1 exclusively or in isolation is to teach something other than science; it converts a science lesson into a lesson in the humanities or the social studies. It is the knowledge and methodologies of science, presented in association with related social and aesthetic values, that constitute human-centered science.

This is particularly essential in connection with the aesthetic subcategories in the category *orientation to the natural environment*. The concepts included as aesthetic subcategories were chosen not only because of the sense of emotional satisfaction created through being aware of them and finding instances of them in the environment, but also because they have clear relevance to conceptualizing scientific principles. These subcategories should be used with this dual purpose in mind. A poet may create a composition whose only purpose is to evoke an emotional response. This is well and good, and is sufficient reason for its existence. In teaching science, however, we must clearly recognize that the purpose of scientific inquiry is to create rational, empirically verifiable explanations of sensory experience. Therefore, when for example a student sees symmetry in a natural phenomenon, or in a representation of phenomena, one should ask what relevance the presence of symmetry has to better understanding of the event or object observed or represented.

Chart 2 presents some examples of scientific significance in aesthetic properties, mainly organized in terms of such functions as stability, efficiency, and specificity. Teachers and students might find it interesting to make similar charts, entering instances they themselves can find. A few additional comments are offered here about the significance of form in physical and biological systems and in their symbolic representations.

Concentricity as observed in a Bohr model of the atom is not merely an interesting geometric pattern but is an integral part of Bohr's concept of atomic function. The concentric orbits were a way to explain the various energy levels of electrons in an atom. Through electron transitions between the orbits, energy absorption and release could be explained. By making assumptions about the filling of the orbits with electrons, certain chemical combining properties of elements were explained. Therefore, when concentricity is perceived in other physical systems or their representations, its scientific significance should be sought, in addition to the aesthetic satisfac-

CHART 2
Scientific Significance of Aesthetic Properties

Category	Significance
Symmetry	*Specificity:* Symmetrically arranged parts provide a particular pattern that through its redundant organization yields specificity or a degree of uniqueness. This enhances facilitatory interaction between appropriate organisms and allows specific combinations of elements and forms in physical systems.
Complementarity	*Recognition specificity:* Examples are enzyme-substrate, antigen-antibody, cellular aggregation reactions, chemical reaction specificity.
	Stability: Complementary relations lead to strengthened associations, interlocking plant and animal communities, signal transmitter and receiver resonance.
Balance	*Stability:* Balance can provide physical stability of form and steady-state stability in functions.
Periodicity	*Efficiency:* An elaborate spatial organization can be produced by repetitive addition of identical units; and temporal patterns can be sustained through repetitive actions (heart pulsation, planetary motion, and standing waves in physical systems).
	Specificity: Redundant patterns contain specificity through repeated content.
	Transmission of energy and information: Waves represent energy transmission through space; and periodic nerve discharges conduct information within living systems.
Concentricity and reticula	*Stability:* Concentric and reticulated forms are resistant to perturbation.
	Efficiency: Packing of material in concentric layers is an efficient form distribution; and reticulate forms allow economical distribution of small amounts of matter over a large surface area (as in nets).
Pattern-unit relation	*Unity, organization:* Recognizing that several particular instances or components may be categorized or organized according to a certain pattern (form or concept) enhances their meaning. For example, the concept of "field" in physics subsumes magnetic, electrostatic, gravitational, and atomic forces that otherwise might be seen as separate and unrelated.

tion it may produce. Similarly, symmetry and balance should be considered in terms of the stability they represent in physical systems. There is very often more order in a symmetrical system than in a non-symmetrical one. The spiral form of some galaxies is explained by one theory as resulting from differential effects of rotation on the astronomical particles (stars) within the galaxy.

Many forms observed in biological systems have survival value for the organism and can thus be related to Darwinian theory of evolution or to explanations of function related to the particular form observed. Spirals in tubular elements such as conducting vessels in plants serve to strengthen the element against collapse. Reticula as observed in leaf venation, circulatory systems in animals, or tissue organization as in sponges provide increased surface area to facilitate exchange of fluids and other materials. Symmetry and balance provide stability in biological systems. Concentricity as observed in bone and in bulbs of plants is an economical way of depositing biologically significant material in a pattern that allows for growth and successive deposition.

It is of course also possible for primary unit properties such as color and sound to have simultaneous aesthetic and scientific significance. The color of an organism while arousing an aesthetic response may also have survival value for the organism by making it less visible in its environment. Animal sounds such as bird calls may arouse an aesthetic response in humans while also serving a communication function for the animals; some bird calls, for example, are used to mark territorial boundaries, reducing the necessity for direct confrontation in defense of territory.

The concept of pattern-unit relation is more general than the other categories we call aesthetic. A *pattern* is any form or concept perceived as ordering or organizing particular elements or phases of phenomena. (Such a concept may be represented by a verbal term, description, definition, law, or principle and/or by a mathematical formula or other non-verbal symbolism.) *Units* so organized may be components of a system or representation, or they may be particular instances of a given kind of organization or relationship. For example, symmetry and the other aesthetic categories are themselves patterns (relational concepts); cited units (instances) of each pattern are related and subsumed by the concept, as are units (components) of each instance.

Recognizing instances of patterns has very general importance in science. In Chart 3, a number of patterns often observed are listed with related units (instances) from the biological and physical sciences. These can be used as themes in preparing laboratory experiences. Each of the particular patterns is a scientific generalization that can be used to order one's thinking about unitary instances subsumed within each pattern. Laboratory experiences

CHART 3: Pattern-Unit Relations

Pattern	Biology	Units

Cycle:

Chemical cycles: Photosynthetic dark reaction, photophosphorylation cycle, Krebs cycle, ATP production and breakdown, protein synthesis and degradation (amino acids yield protein which in turn is degraded to amino acids), starch synthesis and degradation, fatty acid synthesis and degradation, polynucleotide synthesis and breakdown, oxidation-reduction cycles as in cytochrome chain elements, lactate-pyruvate interconversion in muscle, nerve membrane potential build-up and discharge upon stimulation, visual purple in retina synthesis and light degradation, oxyhemoglobin-carboxyhemoglobin cycle in RBC.

Cell cycles: Cell division, chromosome duplication, mitosis, gene segregation and recombination, protoplasmic streaming, contractile vacuole action in protozoa, flagellar motion.

Organ cycles: Heart pumping, pulmonary inspiration and expiration, muscle contraction and elongation, digestive system cycles.

Organism cycles: Reproductive cycles, alternation of generations, diurnal metabolic cycles.

Ecosystem cycles: Nutrient cycles, geochemical cycles, carbon dioxide and oxygen cycles, oxidation-reduction cycles (nitrogen fixation and release), population cycles (oscillations in predator-prey population density).

Gradient:

Cellular gradients: Diffusion pressure gradients across cell membrane, cellular differentiation and maturation (sequence of changes in maturation), lysosomal digestion sequence.

Tissue gradients: Embryogenesis, origin and maturation of tissues, gradation of cell types in tissues as in layers of epidermis, and plant stem tissue arrangement in annular rings.

Organismic gradients: Metamorphosis, maturation of an organism, growth curves, blood pressure gradients in circulatory vessels.

Ecological gradients: Succession of forms of life over time; distribution of plants in a geographical range; population densities around a population center; temperature, mineral, and light intensity gradients.

CHART 3 (continued)

Pattern	Biology *(continued)* Units
Coaction:	*Cellular coactions:* Nuclear-cytoplasmic regulation mechanisms, mitochondrial ATP production and ribosomal protein production, enzyme-substrate feedback control, symbiosis, commensalism, predator-prey relations, food webs, organismic-environmental adaptation.

In addition to the foregoing, there are numerous examples of both spatial and temporal patterns — specific to a given biological phenomenon — that the student must abstract as each instance appears. These include particular distributions of organisms in an ecological system, the arrangement of parts in a whole, and some temporal patterns other than those cited above.

Pattern	Physical Sciences Units
Periodic relation:	Pendulum properties, wave production on water surfaces, instances of interaction of electromagnetic waves, tuned electrical circuits, instances of mechanical sympathetic resonance (tuning forks), wave propagation in coil springs ("Slinky") and in elastic strings, instances of energy related to frequency of radiation $(E = hf)$, wave models of electron distribution in an atom, and astrophysical events such as planetary motion.
Equivalence relation:	Instances of phenomena where one variable is equivalent to another. Examples: light reflection (angle of incidence is equal to angle of reflection); conservation of energy and momentum (momentum or energy gained by a component in a closed system equals the momentum or energy lost by some components in the system); instances of Newton's third law ("Whenever one body exerts a force on another, the second body exerts a force equal in magnitude and opposite in direction on the first body"); balanced reaction equations in chemistry (number of electrons gained equals number lost); and so forth.
Proportional relation:	Instances of the relations between such variables as force and acceleration, velocity and time, weight and gravitation, kinetic energy and velocity, pressure and volume of a gas, frequency of harmonic motion and the length of the string, current and voltage, and so forth. In chemistry, the combining properties of atoms as related to the patterns of activity derived from the periodic chart and instances of phase and chemical reaction equilibria are included here.

should allow students the opportunity to discover these patterns and to identify instances of them. As a teacher you can do much to help students integrate new observations with prior ones by relating new observations to previously identified patterns. Unit-pattern matching is a very basic skill that can and should be facilitated through laboratory instruction.

Five

LABORATORY EXPERIENCES IN A NEW PERSPECTIVE

Teaching has of course as much to do with practice as with scholarly consideration of those qualities of experience that enhance learning. In this chapter we focus on the practical. The conceptual issues discussed heretofore have been related to a three-dimensional model of science processes. I now present the following materials designed to embody the concepts for actual use in teaching: (1) a numerical classification system for analyzing laboratory experiences in terms of the dimensions and categories of the model; (2) five complete biology lab experiences exemplifying categories of the model and systematically analyzed using the classification scheme; (3) brief descriptions of two sets of published laboratory experiences – some in physics, some in biology – with comments relating them to the categories. It is hoped that these materials will be helpful as a guide to analyzing, creating, and directing experiences in various science disciplines.

A Classification System

The classification scheme presented here is a concise way of ordering and labeling laboratory experiences. It provides a numerical notation with which to classify the various parts of a laboratory experience and thus also to conveniently represent the sequential development of the experience in outline form.

CHART 4
Categories and Codes for Classifying Science Experiences

Dimension 1: Orientation to Natural and Social Environments	Dimension 2: Orientation to Science Process	Dimension 3: Science Process
1.1 Natural environment	**2.1 Sensation**	**3.1 Observation**
Aesthetic perception:	2.1.1 Auditory	3.1.1 Unit perception
1.1.1 Symmetry	2.1.2 Gustatory	3.1.2 Pattern perception
1.1.2 Complementarity	2.1.3 Olfactory	*Pattern processes:*
1.1.3 Balance	2.1.4 Tactile	3.1.3 Spatial tracking
1.1.4 Periodicity	2.1.5 Visual	3.1.4 Grouping
1.1.5 Concentricity		3.1.5 Ordering
1.1.6 Reticula	**2.2 Predisposition**	*Cognition:*
1.1.7 Pattern-unit relation	2.2.1 Awareness of organization	3.1.6 Form perspective
Organization perception:	2.2.2 Confidence to sustain scientific activity	3.1.7 Form transformation
1.1.8 Order-disorder		**3.2 Integration**
	2.2.3 Confidence in capacity to reach a goal	**3.3 Interpretation**
1.2 Social environment	2.2.4 Projection of an ideal	3.3.1 Unit explanation
Value judgments:		3.3.2 Pattern explanation
1.2.1 On social action or human behavior	**2.3 Recognition**	3.3.3 Inferred form
1.2.2 On scientific research quality	**2.4 Operation**	3.3.4 Inferred function or effect
1.2.3 On impact of science on society		3.3.5 Classification
1.2.4 On impact of science or society on natural environment		3.3.6 Correlation
		3.3.7 Causal explanation
		3.4 Prediction
		3.4.1 Pattern extension
		3.4.2 Extrapolation
		3.4.3 Hypothesizing
		3.5 Evaluation
		3.6 Reconstruction
		3.7 Construction

Each category and subcategory in the three-dimensional model is assigned a two- or three-digit numerical code, as shown in Chart 4. The first digit identifies a dimension, the second a category within that dimension, and the third (if any) a subcategory of the category. The digits (integers) in such a code are separated by decimal points. Thus, 1.1.2 represents the second subcategory of the first category (1.1) of Dimension 1. This is the code, in other words, for the aesthetic subcategory *complementarity* of the category called *natural environment orientation* in Dimension 1 – *orientation toward the natural and social environments.* In some cases there are no subcategories in a category, and therefore its two-digit code is used to label science activities (for example, the Dimension 2 category code 2.4, for *operation,* is often used); but if there are subcategories in a category, then a three-digit subcategory code is used.

To produce a complete classification of an experience (or part of one), one can use all three dimensions – three numerical codes can be combined. The codes for the three dimensions are separated by slash marks. Thus, 1.1.2/2.1.5/3.1.2 represents perception of *complementarity in the natural environment,* using a *visual sensory modality,* to gather information through *observation of patterns.*

To represent a sequence of activities within a science experience, three-dimension codes for the various activities are separated by dashes. For example, three successive activities of a lab experience might be coded as follows: 1.1.2/2.1.5/3.1.2 – 1.1.2/2.2.1/3.1.2 – 1.1.8/2.4/3.2.

The following outline describes recommended sequential steps for using the category codes to analyze a lesson.

A. To begin analyzing a laboratory experience, classify its first activity or phase by assigning it a code for each of the three dimensions (in reverse order) – first for Dimension 3, then for Dimension 2, finally for Dimension 1 – as follows:

1. First, identify the Dimension 3 category or subcategory that represents the dominant *science process* involved. For example, the first process used is very often observation (category 3.1); and in such cases one of the seven subcategories of this category (usually 3.1.1 or 3.1.2) will be used as the Dimension 3 code.

2. Next, determine the category or subcategory of Dimension 2 *(orientation to science process)* associated with the science process already coded for the activity. If that process is one of the more general subcategories of observation (3.1.1 or 3.1.2), the activity may need to be coded twice for Dimension 2 – first to specify the sensory modality used (2.1.1, . . . , 2.1.5) and then to identify any higher Dimension 2 categories represented in the activity, either simultaneously or in a continuation or repetition of the activity (see the example below).

3. Finally, assign to the activity the most appropriate subcategory code for Dimension 1 *(orientation to environment)*. Most often this will be one of the subcategories of category 1.1, relating to the *natural* environment (as distinguished from the social environment category 1.2).

Example: Suppose a lab experience begins with an observation task using visual data-gathering about patterns in the natural environment. First the dominant process of Dimension 3 is coded (observation for pattern identification): 3.1.2. To this is added the Dimension 2 code for the sensory modality used in the observation (visual): 2.1.5. We now have two of the three parts of a complete code classification for the activity: 2.1.5/3.1.2. Finally the Dimension 1 code for environment orientation is added (perception of order or patterns in nature): 1.1.8. Thus a complete three-dimensional code for this activity is: 1.1.8/2.1.5/3.1.2. But then suppose that in the course of the activity, or following a first phase of it, experiences are provided for reinforcing this same kind of activity, tending to produce or enhance predisposition to it. Then the next complete code should be 1.1.8/2.2.1/3.1.2, in which the Dimension 2 code for sensory modality (2.1.5) is replaced, in the previous complete code, by the Dimension 2 code for predisposition to awareness of organization (2.2.1).

B. Identify the next phase of the lab experience by determining the very next science process that is presented, and classify this phase as outlined in Step A above, working backwards from Dimension 3 to Dimension 1.

C. Apply this procedure repeatedly until the total laboratory experience has been completely coded.

The purpose in coding a laboratory lesson is to summarize the sequence of events that can occur in the experience. The result of coding is an outline of the plan of the experience. (The code numbers are simply labels and should not be interpreted as having any mathematical properties.) It should be noted that coding can be done at different levels of fineness. You can choose to isolate only the longer segments of a laboratory lesson, coding only the larger blocks of experience. Or, if you want to make a finer analysis, smaller segments can be coded. The degree of fineness will of course depend partly on the richness of the activities in the lesson. If there are only a few major blocks of activities and these have fairly homogeneous content, then you will be constrained by the limited material in the experience. In analyzing the following sample lab lessons, I have chosen to code only their major segments. When you code the lessons, you may find additional categories to be coded or indeed subcategories that you feel should be added to the ones I have cited. The classification scheme as shown in Chart 4 contains only the categories and subcategories I have discussed in connection with my model. It is not intended to be absolute or final. You may want to expand the model and the classification system to better represent your own perspectives on laboratory teaching.

Five Laboratory Experiences in Biology

In this section five complete lab experiences are presented and analyzed according to the classification system outlined in the foregoing section. The presentation of each experience is divided into three parts: (1) *Rationale,* explaining basic purposes and assumptions; (2) *Basic Concepts,* presenting fundamental information in organized terms; and (3) *Laboratory Task.* In each case, the *Laboratory Task* section is meant to be suitable for direct presentation to students and is written accordingly.

Lab Experience I

Cell Shape and Function: Units and Patterns

Rationale

The purpose of this experience is twofold: It provides opportunity for students to (1) identify the properties of individual cells obtained from analogous regions in plant and animal tissue and (2) discover the patterns of cellular organization that are found in tissues of similar kinds. The lesson emphasizes use of observation (3.1), integration (3.2), and interpretation (3.3). The skill of interpretation is developed through use of inferred function (3.3.4) as deduced from observations of individuals cells (3.1.1) and of cell patterns (3.1.2). The categories of predisposition (2.2) and perception of organization in the natural environment (1.1.8) are combined to yield experiences that enhance student tendencies to anticipate and seek orderly patterns in their interpretation of experience. This is accomplished by arranging the experience so that students can discover that cells from certain surface layers (epithelia) of an organism, whether animal or plant, have a characteristic structure that is different from most cells found within deeper or internal tissues. The student, with some guidance from the teacher, can find instances of complementarity (1.1.2) – as for example in the spatial relations of epidermal cells in a leaf when the boundary of one cell interlocks with the boundary of an adjacent cell. These sometimes have the kind of complementarity exhibited in the fit of jigsaw puzzle pieces.

In terms of the numerical taxonomy, this complete laboratory experience could be assigned the following code: 1.1.2/2.1.5/3.1.1 – 1.1.8/ 2.2.1/3.1.2 – 1.1.2/2.4/3.2 – 1.1.7/2.4/3.3.2 – 1.1.7/2.4/3.3.4.

Basic Concepts

Animal epithelial tissue is classified as of two kinds: (1) simple epithelium – containing a single layer of cells; and (2) stratified epithelium – containing several layers of cells. In this experience, the students will examine single layers of cells, and therefore the pertinent kinds of simple epithelial cells need to be explained. Simple epithelia are composed of four kinds of cells: (1) squamous, (2) cuboidal, (3) columnar, and (4) pseudostratified. Squamous cells are flat, "tile-shaped" cells. They are thin and sometimes form interlocking patterns. Cuboidal cells are cube-shaped. These cells lie in close association with one another, but usually do not form interdigitating patterns. Columnar cells are tall, with a height considerably greater than their width. Columnar cells form closely packed layers that line surfaces such as the lumen of the intestine. Palisade cells are parenchyma cells in leaves that often appear columnar in shape. Pseudostratified epithelia are formed from a single layer of columnar cells that are staggered in arrangement such that a cross-section through the layer intersects the top portion of some cells and the bottom portion of other cells, which gives an illusion of stratification.

Observation of the interlocking pattern of epithelial cells points toward a function of strengthening this covering layer.

Glands and other soft tissue are composed of polyhedral-shaped cells whose arrangement in the tissue facilitates their function. As an example, the secretory unit in pancreatic acinar tissue exhibits a rosette of cells surrounding a central secretory space. The efficiency of collecting secretory products is clearly deduced from this cell pattern.

Laboratory Task

All living things are composed of individual units of living matter called cells. Some organisms have only one cell. These are sometimes called unicellular organisms. They are abundant in ponds and streams. Among the unicellular organisms are paramecia, amoebas, and other microscopic organisms. Other forms of life have many cells – these are called multicellular organisms. Multicellular organisms have many different kinds of cells, each kind having a particular shape and serving a particular function. The cells in multicellular organisms have become specialized to serve a particular function in keeping the total organism alive. Sometimes we can discover the function of a cell by observing where it is located in an organism and carefully observing its shape and its relationship to other cells that surround it.

Cells that are grouped together and have a similar form and function are called a tissue. Among the various kinds of tissues are (1) those that form linings of surfaces of an organism and (2) those that make up the bulk of organs within the body of an organism.

The purpose of this laboratory experience is to examine cells of various tissues obtained from plants and animals. Using information about their shape and location, you can discover the function served by these cells.

1. Obtain some cells from the lining of your mouth. Use a clean spatula or a clean cover-slip to gently scrape the inside of your cheek. Be careful not to press so hard as to cut yourself. The milky deposit that you obtain contains a suspension of cells. Place a drop on a microscope slide, add iodine stain, and place a cover-slip on the specimen. Examine the shape of the cells and, where possible, the organization of cells when two or more are connected.

    Describe the shape of the cells and their relationship to one another.

    From your observation of the cell shape, size, and pattern of multicellular organization, what do you think is the function of these cells?

2. Examine a slide of a tissue section such as a liver or pancreas tissue preparation. Examine the shape and arrangement of the cells. How does the cell shape and arrangement differ from what you observed in step 1? What would you guess is the function of the cells as inferred from their organization?

3. Obtain a leaf from a broad-leaf plant such as coleus, geranium, or other common household plant. Gently peel away the lining of the lower surface of the leaf. This can be done by tearing the leaf at its margin and then gently pulling the epidermis away from the underlying tissue. Usually a thin portion of the lining will be exposed at the edge of the tear. Place the sample on a slide and prepare a water mount.

    What are the shapes of the various cells you observe?

    How are the cells organized? What patterns can you find?

    From your observation of the shape of the cells, what would you deduce about their function?

    Why do some of the cells surround a narrow opening?

4. Fold a leaf near a vein. Using a razor blade, make several very thin slices (cross-section) through the leaf vein and surrounding leaf blade. Examine some of the thinnest sections with the light microscope, using a water mount preparation.

    How do the cells inside the leaf vein and leaf blade differ from those on the leaf surface as observed in step 3?

    Describe the pattern of the cells in the vein and blade. What do you suppose is their function in each place, based on your observations?

5. Combine your data for plant and animal cells. What generalizations can you make about cell shape and organization in tissues that form linings

in animals and plants? What generalization can you make about cell shape and organization in tissues that are found inside plant and animal bodies?

What precautions must you use in making these generalizations? Do you have enough instances in your samples to say what occurs in all animal and plant tissues of the kind you have observed?

Lab Experience II

Staining Plant Tissues: Pattern-Unit Relations

Rationale

This experience provides opportunity for students to use observation toward discovery of pattern-unit relations. In the first phase, the student acquires information about specific color reactions produced by a metachromatic stain when applied to substances commonly found in plant tissues. This unitary information is then augmented by examination of macerated wood. The kinds of cells in the macerated wood are identified and their color reactions with the metachromatic stain are observed. Their chemical composition is thus determined. As a final step, thin free-hand sections of plant stems are examined, to determine the color reaction of stem tissues and to relate the patterns of tissue organization and staining properties to the unitary information gained in the first two phases.

This laboratory experience may require two periods. The first period can be devoted to phases 1 and 2, and the second to examination of stem cross-sections (phase 3). This experience places emphasis on (1) perception of units and patterns, (2) integration of information through application of the color reactions to subsequently observed cells and tissues, and (3) interpretation of pattern-unit relations through an analysis of the organization of cells in tissue systems.

The three phases of this experience are classified as follows: 1.1.7/ 2.1.5/3.1.1 − 1.1.7/2.1.5/3.1.2/3.2 − 1.1.7/2.4/3.3.2.

Basic Concepts

The metachromatic stain used in these experiences is toluidine blue O. This stain has the remarkable capacity to produce various colors upon reaction with cellular components. The chemistry of the color reaction is not fully understood. Cellulose stains a blue color that is almost identical to the fundamental color of the stain itself; thus there is no particular metachromatic effect. Lignin produces a greenish-blue-to-intense-blue color reaction. Pectin yields a reddish purple reaction, and protein and cell sap produce a

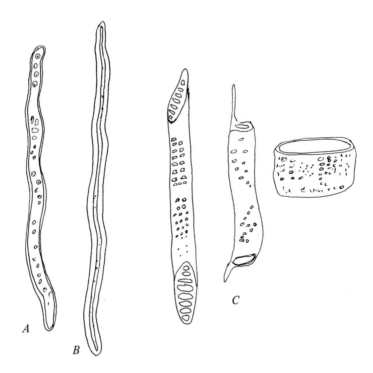

Figure 7. **Cell types in macerated wood.** *A,* tracheid; *B,* wood fiber; *C,* vessel elements.

light purple color. Some of the kinds of cells obtained from macerated wood are shown in Figure 7. These will stain a bright blue if they contain abundant lignin; they give a light purple reaction if they contain largely cellulose and pectin. The color reaction in tissues of young woody stems or new branches of woody trees is most remarkable. The wood stains bluish, the phloem pink-purple, the phloem fibers a deep-blue-to-greenish-blue, and collenchyma a distinct purple color. The collenchyma contains such an abundance of pectin that its reaction is largely purple in color. Herbaceous stems are useful to show vascular bundle organization; older herbaceous stems are more useful for this than younger ones.

Laboratory Task

The cells of plants and animals are made of many different chemical constituents. Plant cells are particularly interesting to examine because they have such varied shapes and differences in chemical content. The cell wall, a unique feature of plant cells, has several different chemical compounds that we can identify readily in the laboratory.

You will be given a blue stain that changes color when it is added to some chemical compounds. When you have determined the color given for each compound, you will be able to apply that information to the analysis of various kinds of plant cells and tissues to determine their chemical composition. You will also be able to determine the pattern of organization of cells in plant stems by using the information gained in the initial steps.

1. Compile a color reaction chart for toluidine blue O, as shown below. Prepare a water mount of each substance. Place the substance on a slide, add a drop of the stain, and allow it to act for about 30 seconds. Place a cover-slip on the preparation and add a little water if needed to completely fill the space under the cover-slip. Observe the color reaction with low power.

Color Reaction Chart:
Toluidine Blue O

Substance	Color
Cellulose (cotton fiber)	_____
Pectin (crystalline)	_____
Lignin (oak wood sliver)	_____
Starch (cooking starch powder)	_____

2. Select a few of the macerated wood samples supplied in the laboratory from various kinds of trees and shrubs. The samples of wood have been treated with an acid mixture to make the cells separate from one another. Take a small portion of a sample, about the size of a pencil lead. Place it on the slide and add a drop of water to rinse the sample. Gently draw the water off by capillary attraction to a piece of filter paper. Add a drop of toluidine blue stain to the preparation and place the cover slip on the slide. Gradually replace some of the excess stain with water by adding a drop of water to one side of the cover slip and drawing the solution off at the other side with a piece of absorbent paper (filter paper). Gently compress and disperse the wood sample by pressing on the cover slip with the blunt end of a pencil or wood stick.

    Observe and describe the kinds of cells and their relationships in the wood from various sources. What color reactions do you observe? Can you make a judgment about the chemical composition of the cells based on the colors observed in Step 1?

3. Obtain stems from some herbaceous and woody plants. Use a fresh razor blade that has been thoroughly washed to make very thin cross-sections of the stems. Practice taking sections until you obtain very thin ones.

Use a toothpick to transfer a thin section of a stem on to a glass slide. Draw off the excess water by use of absorbent paper. Apply a drop of toluidine stain to the section. After 30 seconds, apply a cover-slip and clear away the stain by addition of water from one side. Observe the color reactions of the various tissues in the stem. Can you make a judgment about the chemical composition of each kind of tissue? What is the pattern of organization of tissue in the stem? Based on your observation of color reaction and organization of tissue in the stem, what is the function of each kind of tissue? Identify the woody conducting tissue, the spongy protective and storage tissue, and the phloem tissue.

Compare the various patterns observed in different kinds of stems. What are the differences between the patterns?

Lab Experience III

Algae Classification: Interpretation of Patterns

Rationale

One of the basic processes of science is developing classificatory schemes to organize observations. The production of a classification key is a useful way to introduce students to systematic ways of thinking about organizing experience. In the laboratory experience presented here, moreover, the student is given an opportunity to use aesthetic characteristics of plants as a means of organizing data for production of a key.

The algae are particularly useful as subjects to be systematically ordered using such aesthetic qualities as symmetry, complementarity, balance, periodicity, concentricity, spirals, and reticular organization. In addition, one can attend to more traditional taxonomic characteristics such as cell shape, motility, color, and presence or absence of envelopes surrounding the organism.

The major thrust of this experience is to develop organized ways of interpreting sensory data that have aesthetic dimensions.

A numerical code for the first part of this laboratory experience is 1.1.1-6/2.1.5/3.1.2-5. (In Dimension 1, subcategories 1.1.1 through 1.1.6 are being used, and in Dimension 3 the *observation* subcategories 3.1.2 through 3.1.5 are used.) These same activities are also coded 1.1.1-6/2.2.1/3.1.2-5. *(Predisposition* to be aware of organization should be a product of the observation processes; therefore the code 2.2.1 is appropriate.) In the second half of the experience, students actually produce a classification key. This constitutes *interpretation* by *classification,* at the *operational* level; so the activity is coded 1.1.1-6/2.4/3.3.5.

Basic Concepts

To produce a scientific classification key, one must carefully describe the organisms to be keyed in terms of the characteristics used in developing the key. In this case, the students are to describe algae in terms of the following attributes: symmetry (cellular organization with components in mirror-image design), complementarity (cellular organization where one cell is interlocked with another or some part of the cell exhibits this relationship), periodicity (repeating units as in filamentous multicellular forms), concentricity (concentric layers as in gelatinous envelope layers or in ornamentation of the cell wall), spirals (filament shape and chloroplast shape in some cases), and reticula (net-like organizations holding cells together). In addition, the students should note whether the algae are motile or non-motile, unicellular or multicellular, filamentous or colonial, and blue-green or yellow-green in color.

To begin making a key, each organism should be examined under the microscope and each of the above characteristics when present should be noted on a file card for that organism.

When all of the organisms have been examined and described, they can be grouped, in various ways, into categories based on common characteristics. A binary key is the simplest to produce: the organisms are split into two groups at each decision point. The first split should separate the organisms into two major divisions, if possible based on a salient characteristic present in each group. Then each group is further subdivided into two groups based on those characteristics that clearly separate them. When all organisms have been so split into small groups until each one has been uniquely isolated, the categories are organized as a key. Examples of binary keys are readily obtained in paperback field guides to living things.

Some algae that are particularly useful in preparing keys using the attributes stated above are species of *Chlamydomonas, Cosmarium, Euglena, Gloeocapsa, Hydrodictyon, Oedogonium, Oscillatoria, Pandorina, Spirogyra,* and *Volvox.* If all of these are available, the students should have enough to make a key.

An example of a key to these organisms is given to help you better understand how the aesthetic dimensions are used. Students can of course discover different ways of producing a key. Hopefully, there will be considerable variation among those produced by any class.

1. Alga swims (motile) . . . . 2.
1. Alga does not swim (non-motile). . . . . 5.

2. Alga single-celled . . . . 3.
2. Alga with many cells. . . . . 4.

3. Alga with a large green chloroplast. . . . . Chlamydomonas.
3. Alga with small green chloroplasts. . . . . Euglena.

4. Alga with a net-like mesh (reticulum) connecting the cells . Volvox.
4. Alga with a radial symmetry, but no net-like connections. . Pandorina.

5. Cells in a periodic or repeating pattern in a line . . . . 6.
5. Cells in clumps or single cells . . . . 8.

6. Chloroplast spiral-shaped. . . . . Spirogyra.
6. Chloroplast various shapes . . . . 7.

7. Cells bright yellow-green . . . . Oedogonium.
7. Cells blue-green . . . . Oscillatoria.

8. Cells with symmetrical halves . . . . Cosmarium.
8. Cells arranged as clumps or nets . . . . 9.

9. Cells with concentric layers of gelatin around them. . . . . . Gloeocapsa.
9. Cells forming a net . . . . Hydrodictyon.

Laboratory Task

One of the tasks of a scientist is to discover orderly ways of organzing data. Living things can be described and organized so that other scientists can identify examples of them.

You will be given a set of algae. The name of each alga is on its tube. Use the list of characteristics provided by the teacher and any other ones you can see when examining the organism under the microscope to describe each alga. Your teacher will tell you how to keep a record of what you see.

You will also be told the general principles of how a scientist makes a key or check-list to identify living things. Then you will compose a key of your own, classifying the algae you have observed and described in terms of their distinctive and shared characteristics.

Try to be as accurate as you can in describing the algae and in organizing your data in preparation for writing your key.

Lab Experience IV

Animal Behavior: Model Construction

Rationale

A reasonable attempt is made in this experience to introduce students to the operation of constructing a model of animal behavior that is sufficiently sophisticated to qualify as a "small theory." It is an open-ended experience that allows the students to use certain basic data-collection devices such as observation charts and simple notation schemes toward synthesis of an explanation of interrelationships among animal behavior patterns. The model should be sufficiently general that the students can make predictions about behavior patterns and can evaluate the predictions for their validity through further observation. The lesson provides opportunity for experience in stating both conceptual and operational definitions of the behavior patterns observed.

The students are presented with three small aquaria (or one-gallon display bowls) containing respectively (1) two male fish, (2) two female fish, and (3) a male and a gravid female fish (preferably guppies or swordtails). This laboratory experience begins with direct observation of patterns of fish behavior: 1.1.8/2.1.5/3.1.2. During observation, the student must be able to keep a mental orientation as to left and right sides of the fish. Since there is considerable swift movement during interaction between the fish, it is necessary to represent left side and right side mentally. This cognitive function is what we call form transformation, and its operation in this activity is coded 1.1.8/2.4/3.1.7. Thereafter, the students collate information into a summary of frequencies of behavior (1.1.8/2.1.5/3.1.2). From these data, the students generate an explanation of fish behavior that includes the dominant forms of behavior and their probable role in fish interaction. This is an instance of simple theory construction and is classified 1.1.8/2.4/3.7. There can be a step preceding this one, in which the students perform a logical analysis to extend patterns previously observed (1.1.8/2.4/3.4.1). Finally, if on subsequent occasions the students choose to evaluate the reliability of their explanations through further observation and validation of behaviors observed, that sequence of activities is classified 1.1.8/2.1.5/3.1.2 − 1.1.8/2.4/3.5.

Basic Concepts

Small tropical fish provide interesting repertoires of behavior that can be examined with a fair degree of reproducibility. Many of the viviparous fresh-water forms are useful. Among these are the guppies and swordtails. The guppies are particularly active and can be obtained as new large forms with colorful pigmentation.

Chart 5 presents a classification that can be used in various ways to summarize animal orientation activity. Some possibilities for using it in this lab experience will be suggested here. Some students may see ways to modify or refine the system, or may want to devise their own notation for behavior patterns; if so, they should be encouraged and helped to do so.

The chart can be used in this experiment to facilitate observation and recording of a limited number of behavior interactions between two fish of either the same or different gender. Where both fish are the same gender, they should be of different sizes to allow identification of each. The chart is

CHART 5: Animal Interaction Record Sheet

Organism B: Body Part Approached	Organism A Activity: Body Part Presented and Orientation				
	Head		Tail		Side
	Direct	Angle	Direct	Angle	
HEAD: Front					
Dorsal					
Ventral	– –				
Left Side					
Right Side					
BODY: Dorsal					
Ventral					
Left Side	+ + +				
Right Side					
TAIL: Back					
Dorsal					
Ventral					
Left Side					
Right Side					

especially useful for recording directed behavior of male fish toward female fish. For this purpose, the columns under the heading *Organism A Activity* refer to aspects of male fish behavior and the row categories listed under *Organism B* represent parts of the female fish's body toward which the male orients.

The orientation column categories represent the position of fish A relative to fish B when A is approaching or hovering near B: *Direct* means that A proceeds with its body axis almost perpendicular to B; *Angle* means that A hovers or moves at an oblique angle to the long axis of B. The column categories *Head, Tail,* and *Side* denote which part of organism A is presented toward B. *Head* means that A approaches B head first; *Tail* means it backs up toward B; and *Side* means it hovers parallel or almost parallel to the body of B (therefore the direct/angle distinction is omitted under this category).

The row categories specify which part of organism B is the object of orientation by organism A, and the direction of approach. The part approached can be *Head, Body,* or *Tail.* Under the category *Head,* the element *Front* means a mouth-to-mouth orientation. *Dorsal, Ventral, Left Side,* and *Right Side* mean respectively that organism B is approached from above, from below, from left, or from right. These elements have the same meaning in the *Body* and *Tail* categories. However, in the category *Tail* there is the additional element *Back.* This means that A approaches B from behind.

Many of the behaviors observed will result in a reaction of attraction or repulsion. An attraction occurs if approach behavior is accepted by standing still or by moving toward the approaching organism. Repulsion occurs when organism B darts away (or when both fish do) following an approach. When marking the chart, a plus sign (+) is entered in the appropriate square when the interaction is attractive and a minus sign (–) is recorded when the interaction is repulsive.

To mark the chart, one first notes an action of organism A and identifies the column representing the part of A's body presented to B and the orientation. Then the row representing the part of B's body approached is identified, and the proper square for tallying is identified where the row and column intersect. If the interaction produces attraction, a plus is entered in the square. A minus is entered if repulsion occurs. Each time the same behavior pattern is discerned, it is marked the same way in the same square. In Chart 5, five instances of B's response to A's approach behavior are recorded to illustrate the method. Three plus signs are entered in the square under *Head/ Direct* (organism A) and across from *Body/Left Side* (organism B). This means that three times the male fish (A) approached the female (B) and hovered briefly with his head pointed toward her left side and with his body at right angles to hers. The plus signs mean that the female accepted each approach behavior. Two minus signs are entered in the square under *Head/*

Direct and across from *Head/Ventral*. This indicates that A twice approached B from beneath the mouth and that B darted away both times.

If the students are unable to deal effectively with the complexity of the chart, it is possible to assign parts of the chart to different members in the group. Thus, one person watches only for approaches of A toward the head of B, another for approaches toward the body, and another for approaches toward the tail. This reduces the complexity of cues that each student must attend to in making observations. The students should familiarize themselves with the chart and spend some time in making preliminary observations before marking squares. Observation and recording periods of fifteen to twenty minutes should yield enough tallies to begin to see patterns of responding.

After collecting sufficient data, the students may want to add narrative descriptions about fin flutter and spreading exhibited with some of the categories marked in the chart. The final collated data should be used to write a description of the dominant behavior patterns observed and the differences or similarities observed in patterns between male-male and female-female interactions as compared to male-female interactions. The predictions about dominant behavior patterns can be evaluated on future occasions for consistency.

Laboratory Task

The three small aquaria you are to observe contain pairs of fish. One contains two male fish, another two female fish, and a third contains a male and female fish. All of the fish belong to the same species. From your knowledge about coloration in birds or perhaps in other kinds of animals, which fish do you think are male and which are female?

Now watch the fish in each of the aquaria in turn. It is interesting to determine how the two fish react to one another. How does each one respond to approach from the other? What regular patterns can be discovered? What, if any, are the most frequent patterns of behavior and what evidence can you gather to determine the effect of these behaviors in the interaction between the two fish? Write an explanation of your findings as to the dominant forms of behavior and the probable communication value they have in fish interaction.

Your teacher will give you some advice on how to begin making observations. Then, use your best skill at observing and recording data, and your careful application of imagination, to produce an explanation for the patterns of behavior that you observed. If you have time on another day, you can make more observations to see if your explanations are still valid. This will serve as a check for the generality of your findings.

Lab Experience V

Simulated Cell Membranes: Evaluating Predictions[1]

Rationale

Modern cell biology has made considerable gains through the use of model systems. In this laboratory experience, students are given the opportunity to perform simulation experiments and, from their observation of these systems, to make predictions about lipid activity in cell membrane permeability. The predictions are of a high-order kind approaching hypothetical statements, since the students identify relationships between two variables and evaluate their predicted relation by doing further experiments.

The fisrt experiment provides an opportunity for students to observe the spreading effect of certain lipids on a water surface. This is classified as observation (1.1.8/2.1.5/3.1.2). The second part of this experiment, on the viscosity of lipid films, requires that the student make a prediction about the probable permeability of various lipids to water. This is classified as prediction (1.1.8/2.4/3.4.1); its operation means that the student extends a pattern previously observed to include new instances. This prediction is evaluated in the evaporation experiment (1.1.8/2.4/3.5). Finally, the last phase asks the student to make an interpretation of relationships among the data he has collected (1.1.8/2.4/3.3.2).

Basic Concepts

The purpose of these experiments is to demonstrate the possible role of lipids in regulating molecular diffusion across a cell boundary. The experiments are arranged in a sequence that allows the students to make hypotheses, which are treated in succeeding experiments.

Essential concepts that should be acquired during the laboratory experience, to facilitate student formulation of hypotheses, are the following:

1. Polar lipids such as stearic acid and cholesterol, containing a charged group, when placed on water will orient at the interface with their polar groups in the water phase and their fatty portion projecting away from the water surface.

2. Stearic acid molecules bearing a long-chain fatty tail can pack closely together at an interface, being stabilized by weak intermolecular attractive forces.

3. Cholesterol molecules containing ring structures cannot pack as closely together as stearic acid molecules. (Using concepts 2 and 3,

[1] O. R. Anderson, "Experiments with the Role of Lipids in Cell-Membrane Permeability," *The American Biology Teacher, 32* (1970), 154.

ask the students to predict which lipid would form a solid or immovable film on a water surface.)

4. Cell membranes may contain a lipid bilayer laminated between two protein layers. The internal lipid bilayer may influence the rate and kind of molecules diffusing across cell membranes.

The three experiments to be presented are concerned with (1) producing lipid films on water surfaces and determining their viscosity (i.e., rigidity); (2) measuring the rate of water diffusion across lipid films; (3) assessing the rate of cation diffusion across lipid films at an n-butanol-water interface.

The students, with the assistance of the instructor, should extrapolate their findings to predict the role of lipids in cell-membrance diffusion processes.

Laboratory Task

Experiment 1. Formation of lipid films on a water surface.

Purpose: To determine the viscosity of lipid films.

Procedure: Prepare separate solutions of 0.1g stearic acid in 100ml hexane and 0.6g cholesterol in 100ml hexane. The cholesterol solution may need to be shaken vigorously to achieve complete solubility. Warm the test tube by enclosing it within the hand to increase solubility. Place two petri dishes on a black surface and fill them almost full with clear water. Gently blow some Lycopodium powder on the surface of the water: this can be done best by blowing gently on a small amount of powder in the palm of the hand held at an angle close to the water. Gently add one drop of stearic acid solution on the water surface and note how the powder is dispersed by the lipid film.

What is the area of the lipid film?

Continue adding drops of stearic acid solution until the powder is completely spread to the periphery of the dish. One additional drop should be added to produce a well packed film. The last drop added should not spread completely, thereby indicating that a fully packed film has been formed.

How many drops were required to reach this point? These data should be recorded for future reference when films will be cast without the aid of powder.

Prepare a cholesterol film in the second petri dish using the same procedure as with the stearic acid solution. Record how may drops were needed to produce a complete film.

Now, using clean and labeled petri dishes containing water, cast a film of cholesterol on the water surface of one dish and a film of stearic acid on the other. Gently blow some light talc on the surface of each film. Cautiously puff air onto each surface and determine which of the films is the more fluid.

(A nonviscous, or fluid, film will allow the talc to move around on the surface when it is blown.)

Which film is solid? What does the state of the film (solid or liquid) tell you about the packing of the molecules in the film? Which of the films would allow most diffusion of molecules through it?

Experiment 2. Testing the water permeability of lipid films.

Purpose: To demonstrate the relative water permeability of cholesterol films and stearic acid films, thereby demonstrating the effectiveness of these lipids in regulating the passage of water through cell membranes.

Procedure: Use the same cholesterol and stearic acid solutions as were prepared in Experiment 1. Place three petri dishes of the same diameter, marked A,B, and C, in a secluded place away from dust and drafts. Now place 50ml of clear water in each dish. Be very careful to measure the water accurately, by using the same graduated cylinder for each measurement. Produce a monolayer of cholesterol on the water surface in dish A and a stearic acid monolayer in dish B. Dish C will serve as a control and contains no lipid film on its surface.

After 24 to 48 hours, measure the volume of water remaining in each petri dish by using a graduated cylinder. (Be sure to dry out the cylinder between measurements.) Make a table showing the amount of water lost in each dish.

How can the observed differences be explained? Were your predictions confirmed about the permeability of different lipids to water?

Experiment 3.   Selective ion permeability of lipid monolayers formed at an n-butanol-water interface.

Purpose: To demonstrate that cholesterol and stearic acid monolayers differ in their ion permeability.

Procedure: Fill a test tube one-half full with water lightly tinted with food coloring. Gently add a layer of n-butanol on the water surface.

Where does the butanol remain in the test tube?

Gently rock the test tube and note that a distinct surface or interface has formed between the two liquids.

Prepare a solution containing 0.4g calcium chloride in 100ml of n-butanol. Crush and stir the calcium chloride vigorously to obtain a good solution. Filter or centrifuge out any remaining solid. Prepare three separate solutions from this stock solution, as follows:

Solution A: To 10ml of butanol-calcium solution add 0.1g cholesterol.

Solution B: To 10ml of butanol-calcium solution add 0.1g stearic acid.

Solution C: Reserve 10ml of butanol-calcium solution to be a control.

The calcium ions in each of these solutions will act as marker ions to determine their rate of diffusion through a butanol-water interface.

Now prepare three test-tubes, marked A, B, and C, containing 10ml of 1% aqueous sodium carbonate. The carbonate will react with the calcium to form a white calcium-carbonate precipitate, indicating the presence of calcium diffusing through the interface.

The butanol solutions A, B, and C should be added to their corresponding water-containing test-tubes: be certain that the labels on the test-tubes match. Each solution should be simultaneously trickled down the inner side of its tube to form a stable interface. Immediately observe the test tubes and note when a white precipitate begins to form in each of the tubes.

Where is the most precipitate formed and where the least? What does this imply about the permeability of different lipids to ions? How does the ion permeability of lipid films correspond to their water permeability?

Some Published Laboratory Experiences

Many modern laboratory experiences published by national curriculum committees such as BSCS (Biological Sciences Curriculum Committee), P.S.S.C. (Physical Sciences Study Committee), Harvard Project Physics, and Chem Study can be adapted to the new perspectives presented in this book. To facilitate the incorporation of such materials into our context, some examples of curriculum experiences in two of these programs are briefly described and discussed here, and classified using our numerical scheme.

Biological Science

First we look at seven lab experiences in *Biological Science — An Inquiry into Life: Student Laboratory Guide.*[2] Each is cited by title and number as presented in the Guide; page numbers are also indicated.

Life in Unexpected Places: Inquiry 1-1(p.1). This investigation directs the student toward considering differences between living things and non-living things, and also the origins of living things. In the first phase of this investigation, the students examine pond water, soil, and dry grass for evidence of living things. This suggests classification as 1.1.8/2.1.5/3.1.2. The students mainly use visual observation. In the second phase, the students begin a simple experiment to determine the effects of various materials added to bread. Since the students do not make a prediction about what will happen, but are

[2.] American Institute of Biological Sciences, *Biological Science — An Inquiry into Life: Student Laboratory Guide* (New York: Harcourt, Brace and World, 1968).

only asked to interpret what happens, this is categorized as simple controlled experimentation; the classification is 1.1.7/2.4/3.3.7. The last code indicates causal relations. This is used because the students are to determine which materials, when added to the bread, give rise to living things.

The final phase of this experience provides for multiple-trial experimentation. The students heat inoculated specimens to various temperatures to find out which temperature and duration of treatment produces sterilization. This is classified as 1.1.8/2.4/3.3.4. The last category indicates that the students are to infer the effects of their treatment in killing organisms by examining the test tubes for indirect evidence of life.

This investigation begins the laboratory experiences in this curriculum. It is clear, by examination of the classification categories, that the exercises involve the simpler categories of our model.

The investigation could be further adapted to the rationale of this book by including some discussion of man's belief about the importance of controlling life, as in medical applications of control of disease vectors and pathogens. What assumptions are made about the importance of human life? Such discussions fall under category 1.2.1 — value judgments about social action. They could also include topics on the impact of science on society through control of pathogens. The extended lifetime of modern man and the effects of freedom from disease on human productivity could be discussed as products of applied science.

Life from Nonlife: Inquiry 2-1 (p. 22). This is a continuation of Inquiry 1-1 and extends the data obtained there to allow the students to make predictions about what culture conditions started in that inquiry will produce life. This experience is classified as 1.1.8/2.4/3.4.2. The students are asked to make extrapolations based upon their interpretations. This investigation also allows for development of Dimension 1 categories 1.1.7 and 1.2.3. Category 1.1.7, pattern-unit relations, is suggested because the students have data that will allow them to generalize as to what temperature ranges and durations are sufficient to cause death of organisms. This pattern can be used to recommend particular applications in laboratory research and industry. Category 1.2.3, value judgments about the impact of science on society, could relate to discussions of the contribution of science to the rise of the canning industry, the dependence of modern medicine on asepsis through heat treatment, and man's ability to be more independent of the environment.

It is also clear that this experience can contribute to a predisposition to anticipate organization in perception (2.2.1) and confidence in one's capacity to reach a goal (2.2.3), since the experiment lasted over a period of a week. This experience also includes integration (category 3.2); the students use prior-gained data in making interpretations of the newly collected data.

Cork – An Investigation into Form and Function: Inquiry 3-1 (p. 23). The students examine cork cells and compare them to other material (1.1.8/ 2.1.5/3.1.2). They further examine the cork and generate an explanation of the structure of cork based on the unitary properties observed in the first phase (1.1.7/2.4/3.3.2). In the last part of the investigation, the students are asked to examine the cork for evidence of prior life. This involves inferred function (3.3.4).

Oxidation-Reduction in Living Cells: Inquiry 5-4 (p. 38). In this investigation, a chemical oxidation-reduction reaction is observed and a method of detecting this activity is identified. Then, the same method is applied to yeast cell cultures to detect oxidation-reduction reactions in living systems. This is an instance of finding unit-pattern relations; it is classified as 1.1.7/ 2.4/3.3.2. The evidence for oxidation-reduction at various places in the yeast culture is determined. This constitutes pattern interpretation.

Another example of pattern interpretation and identification of unit-pattern relations occurs in Inquiry 9-2 (p. 69). Here the students develop a dossier of characteristics for three viruses (bacteriophages) and are told they can use this pattern to identify instances of a particular virus.

The Environment of a Microorganism: Inquiry 11-3 (p. 78). The students make observations about temperature of incubation and bacterial growth. The data are then used to plot a graph showing the relation between growth and temperature. This is classified as 1.1.8/2.4/3.3.6. Moreover, the students are asked to irradiate yeast with ultraviolet light for varying periods of time and to graph the relation between radiation dose and growth. This is also classified 1.1.8/2.4/3.3.6. Then, they are asked if it is possible to extrapolate the curve. This requires recognition of a scientific process and is classified as 1.1.8/2.3/3.4.2.

Control of Muscle Contraction: Inquiry 25-1 (p. 162). This investigation illustrates use of a laboratory experience to challenge students to extend patterns they have observed in muscle and gland reactions to chemical stimulants. From data they have gathered on acetylcholine and adrenalin effects on heart action and stomach tissue, the students are asked to extend the data by including the mediating influence of the nervous system in these reactions (1.1.7/2.4/3.4.1). Moreover, the students are invited to formulate a hypothesis about the effects of the drugs on striated skeletal muscle (1.1.8/ 2.4/3.4.3). If time allows, they may test their hypothesis (1.1.8/2.4/3.5).

Inheritance of One-Factor Differences: Inquiry 29-3 (p. 188). This investigation of Drosophila genetics demands a high degree of student involvement in prediction and evaluation. The students make predictions of kinds of offspring expected based on genetic characteristics of adults; this is classified as 1.1.8/2.4/3.4.1. The pattern prediction code (3.4.1) is used because the data are derived from analysis of gene combination patterns. The students test their predictions using chi square statistical methods; this is classified as 1.1.8/2.4/3.5.

The Biological Science Inquiries: General Comments.   The investigations cited above show a clear progression of complexity in the kinds of processes students are invited to explore.   The earliest experiences are largely observation-type experiences with few invitations for the student to recognize the appropriateness of prediction and evaluation processes.   The later experiences progress more and more toward operational use of interpretation, prediction, and evaluation.

It is noteworthy that there is little evidence of direct encouragement to use reconstruction and construction processes. Moreover, there is not a major thrust to develop theory-based experimental research designs, although the student is asked to generate *ad hoc* explanations for some observations. The clear link between theory and hypothesis is not readily discernible in many of the investigations.

Some additional emphasis on evaluation of scientific investigations with respect to social impact (1.2.3/2.4/3.5) could be beneficial. There is some evidence throughout the book of encouragement for students to evaluate the quality of experimental design (1.2.2/2.4/3.5).

PSSC Physics

As examples of modern physical science laboratory investigations, five experiments published in *PSSC Physics: Laboratory Guide*[3] are analyzed here.   Each experiment is cited by title as presented in the laboratory guide; page numbers are also indicated.

Experiment 1 – Analysis of an Experiment (p. 1).   The experiment asks students to consider the relationship between time required for a can of water to empty through a hole in the bottom when the diameter of the hole and the height of the water in the can are varied. The experiment is to be interpreted by the student and rapidly introduces some of the higher-level processes represented in our model.

The student is first asked to plot emptying time versus diameter of opening for a constant height; this is classified as 1.1.8/2.4/3.3.6. The student is then asked to recognize that such mathematical relations can be used to extrapolate trends (1.1.8/2.3/3.4.2). The student is then asked to recog-

3. U. Haber-Schaim and others, *P.S.S.C. Physics: Laboratory Guide* (Lexington, Mass.: Heath, 1971).

nize that a prediction can be made for trying a different kind of plot (1.1.8/ 2.3/3.4.1). The student then evaluates the procedure by actually generating a graph (1.1.8/2.4/3.5). Moreover, the student is asked to make a judgment about reliability of observations (1.2.2/2.4/3.5).

It is interesting to note that this experiment, the first in the *PSSC Physics* program, begins with direct use of advanced processes and orientations that were introduced in the *Biological Science* program only after considerable prior development of less complex categories.

Experiment 2 – Reflection from a Plane Mirror (p. 3). The experiment begins with an observation of points where rays of light reflected from a mirror are intersected (1.1.8/2.4/3.1.2). The category of pattern identification (3.1.2) is involved because the student identifies patterns of light ray reflections. The student then makes an interpretation of the patterns so obtained (1.1.8/2.4/3.3.2).

Experiment 6 – The "Refraction" of Particles (p. 9). This interesting simulation experiment begins with observation of paths that rolling balls take when released down an incline (1.1.8/2.1.5/3.1.1). These unitary observations are then compared to a general principle (pattern), Snell's law, to determine how well they correspond (1.1.7/2.4/3.3.2). The students are then asked if they could make a lens to "focus rolling balls." This recognition task is classified as (1.1.7/2.3/3.4.1).

Experiment 13 – Waves from Two Point Sources (p. 22). This experiment begins with an opportunity for students to identify patterns as a function of frequency of perturbation of a water surface (1.1.8/2.4/3.3.2). The students then use a mathematical formula to make a prediction about the position of a certain pattern and are asked to check the prediction. This sequence is classified 1.1.8/2.4/3.4.1 – 1.1.7/2.4/3.5.

Experiment 24 – Centripetal Forces (p. 44). This is an interesting example of theory-guided research using multiple trials. A theoretical statement about circular motion sets the stage for this experiment. The rate of revolution of a whirling body at the end of a fixed string with varying amounts of tension is measured as a function of the centripetal force produced by the tension. This requires multiple trials; the rate of rotation is repeatedly assessed as more weight is added to the string. The experiment requires explanations relating period of motion to centripetal force (1.1.8/2.4/3.3.6), and predictions based on observations (1.1.8/2.4/3.4.1).

P.S.S.C. Physics: General Comments. This set of experiments exhibits some interesting properties. The first experiment requires rather sophisticated use of higher scientific process categories, but does not lead into reconstruction or construction tasks. The other experiments place emphasis on pattern explanations, finding unit-pattern relations, making predictions, and evaluating predictions. There is some opportunity to perform experiments that are clearly theory-based. One notable consistent omission is any attempt to engage the student in making socially relevant evaluations of the outcomes of the experiments; nor is there an attempt to have the student analyze the attitudes and beliefs that guide his performance of the experiments. Although there are frequent invitations for students to generate more experiments and to give additional explanations for observed phenomena, there is little attempt to have the student analyze the assumptions underlying his actions. Moreover, there are many instances in these experiments where aesthetic categories such as symmetry, complementarity, balance, and periodicity could be more explicitly applied.

The experiments do make a fair amount of use of pattern-unit relational analysis through application of generalizations and mathematical relations to unitary or paricular instances subsumed by them. (Recall that Chart 3, in Chapter Four, provides various examples of pattern-unit relations that can be applied in biological and physical science instruction.)

There are few codable instances, in the P.S.S.C. experiments, of predisposition subcategories (2.2.1-4). These are difficult to detect in written descriptions of laboratory exercises, and apparently curriculum writers in the past have not given careful thought to encouraging student responses of this kind. The category of predisposition, moreover, is one of the most subtle and elusive ones in the classification scheme. Yet it is highly significant for the development of a stable and mature intellectual disposition. Much of this development depends more on the attitude and social pressure (encouragement) initiated by the teacher than on the content of written laboratory instructions. I hope that the attention given to predisposition learning in this book will help teachers to foster more of this kind of tacit learning in their classrooms and labs.

Significance of Predisposition Learning

An additional comment on predisposition is offered here in light of the observations made in this chapter.

The four predisposition subcategories of the model represent three kinds of mental orientation: (1) awareness, (2) confidence, and (3) projection. Awareness is best achieved through repeated exposures to situations requiring apprehension of the phenomena to be assimilated by the student. Thus subcategory 2.2.1 (awareness of *organization* in perception) can be developed only as the student is repeatedly exposed to situations that press him to create rational and orderly explanations for what he observes. Such experiences must come regularly throughout the learning experience of several months, and each approximation of progress toward order perception must be reasonably reinforced.

Confidence subcategories (2.2.2 and 2.2.3) are less easily manipulated. A student's self-esteem and his assurance about his capacity to perform activities is no doubt regulated by numerous psychological variables, including early family life experiences and the extent to which he has been given school tasks of appropriate difficulty and has received adequate reinforcement upon their successful completion. The science laboratory is an excellent setting for encouraging student confidence. Many high school science experiences are a new *kind* of experience for students and, as such, are not liable to arouse feelings of incompetence or failure that may have developed from earlier academic and social encounters. The science teacher is therefore in a very good position to begin to build a student's realistic confidence in his ability to perform intellectual tasks based on empirical data-gathering. It is critically important to seek every opportunity, in the first laboratory experiences of a term, to identify students who seem unconfident. Every realistic evidence of successful approximation to a correct completion of a mental or manipulative step should be quietly recognized and encouragement given for additional progress. This procedure will require more than the normal amount of attention to the general disposition students exhibit in the laboratory. Considering the many other competing requirements for a teacher's attention in a laboratory teaching situation, I know it seems hard when one is asked to attend to an additional perspective: empathy. Yet this is perhaps the most fundamental of the teaching perspectives required for successful guidance of student growth toward higher cognitive levels of achievement.

Projection (2.2.4) is a category that goes far toward enhancing those qualities of experience that clearly set man apart from lower forms of life. It is the ability to project an image of an ideal position one would like to achieve and to orient one's activity toward realization of the ideal. Science instruction can do much to encourage students to broaden their vision of personal and mental development. Many laboratory experiences are of the kind that last for several weeks. Thus the student can be encouraged to set a long-term goal for personal growth and to actively pursue realization of the goal. Students can be encouraged to set goals of improving their psychomotor, cognitive, and group social skills. A student should be encouraged to appreciate his enhanced mental capacity to comprehend the world through enlarged understanding of concepts and principles acquired in science learning. The use of pattern-unit teaching should greatly enhance the student's ability to recognize instances of scientific principles in daily experience. It is very important that students be given the opportunity to identify personal qualities that they choose to refine and opportunity also to set a definite course of action to realize that goal. The emphasis, then, in the category of projection is sensing a higher level of one's capacity and setting a clear course of action toward realization of that capacity.

I hope that the perspective presented in this book will encourage science teachers to seek opportunities of the many kinds suggested — to foster humane growth and intellectual development — and that these will lead to increased satisfaction in the accomplishments of teaching science as a human enterprise.

REFERENCES CITED

American Institute of Biological Sciences. *Biological Science – An Inquiry into Life: Student Laboratory* Guide. New York: Harcourt, Brace and world, 1968.

Anderson, O. R. "Experiments with the Role of Lipids in Cell-Menbrane Permeability." *The American Biology Teacher, 32* (1970), 154.

Anderson, O. R. *Teaching Modern Ideas of Biology.* New York: Teachers College Press, 1972.

Bloom, B. S. (Ed.), and others. *Taxonomy of Educational Objectives, Handbook I: Cognitive Domain.* New York: McKay, 1956.

Guilford, J. P. *The Nature of Human Intelligence.* New York: McGraw-Hill, 1967.

Haber-Schaim, U., and others. *P.S.S.C. Physics: Laboratory Guide.* Lexington, Mass.: Heath, 1971.

Margenau, H. *The Naure of Physical Reality.* New York: McGraw-Hill, 1959.

Meyer, B. J. F., and G. W. McConkie. "What Is Recalled after Hearing a Passage?" *Journal of Educational Psychology, 65* (1973), 109.

Neisser, U. *Cognitive Psychology.* New York: Appleton-Century-Crofts, 1967.

Suchman, J. R. "A Model for the Analysis of Inquiry." In *Analyses of Concept Learning,* ed. by H. J. Klausmeier and C. W. Harris. New York: Academic Press, 1966.

INDEX

Note: In addition to the references indexed here, there are numerous brief references to categories of the model, or to their classification codes, in the laboratory experiences presented or described on pages 105-126. The classification codes are shown on page 102.